GLOBAL WARMING IN LOCAL DISCOURSES

Global Warming in Local Discourses

How Communities around the World Make Sense of Climate Change

Edited by
Michael Brüggemann and
Simone Rödder

OpenBook
Publishers

https://www.openbookpublishers.com

© 2020 Michael Brüggemann and Simone Rödder. Copyright of individual chapters is maintained by the chapters' authors.

Global Communications vol. 1 | ISSN 2634-7245 (Print) | ISSN 2634-7253 (Online)

ISBN Paperback: 9781783749591
ISBN Hardback: 9781783749607
ISBN Digital (PDF): 9781800641259
ISBN Digital ebook (epub): 9781783749386
ISBN Digital ebook (mobi): 9781783749393
ISBN XML: 9781783749409
DOI: 10.11647/OBP.0212

Cover design by Anna Gatti based on a photo by Duangphorn Wiriya on Unsplash at https://unsplash.com/photos/KiMpFTtuuAk

Contents

Acknowledgements

Editing this book would not have been possible without the continuous support from a number of people whom we thank very much. Obviously, the volume would be nothing without the chapter authors' willingness to condense bigger research projects into book chapters and going through several rounds of revisions. We also acknowledge the great support of our student assistant, Joana Kollert, in putting this book together. Thank you to the anonymous reviewers of the individual chapters and overall book concept, and to our colleague in Hamburg, Michael Schnegg, who has provided valuable feedback on the introduction. Kelley Friel has provided support in copy-editing the chapters into better English. We are indebted to Sven Engesser who has taken up the responsibility for this book among the editors of the Global Communications Book Series. Also, we thank Alessandra Tosi from Open Book Publishers who has supported the book and the book series over the years, and who never lost patience with us as the project proceeded slower than expected. We would also like to thank the support team at Open Book Publishers: Adele Kreager, for copy-editing, Anna Gatti, for cover design, and Melissa Purkiss, for typesetting. Finally, we acknowledge funding by the Deutsche Forschungsgemeinschaft (DFG, German Research Foundation) under Germany's Excellence Strategy—EXC 2037 'CLICCS—Climate, Climatic Change, and Society'—Center for Earth System Research and Sustainability (CEN), Universität Hamburg.

Author Biographies

Dorothee Arlt has been teaching and researching at the Institute of Communication and Media Studies at the University of Bern as a Senior Assistant since 2013. Her research focuses on political communication, media in the context of flight and migration, science, energy and climate communication, as well as media reception and impact. Dorothee Arlt studied Applied Media Science at the Technical University of Ilmenau.

Michael Brüggemann is Professor of Communication Research, Climate and Science Communication at Universität Hamburg and Principal Investigator in the Cluster of Excellence CLICCS (Climate, Climatic Change, and Society). His research explores the transformations of journalism, political and science communication from a comparative perspective. For recent publications, see www.bruegge.net. Commentary on climate communication may be found at www.climatematters.de.

Fenja De Silva-Schmidt received her MA in Journalism and Communication Studies at Universität Hamburg, where she is also currently working as a Research Associate to the Chair of Communication Research, Climate and Science Communication. In her PhD dissertation, she analyzes how media coverage and interpersonal communication influence knowledge acquisition about climate politics.

Sara de Wit joined the Institute of Science, Innovation and Society (InSIS), University of Oxford, as a postdoctoral Research Fellow in 2017. She is currently part of the Forecasts for Anticipatory Humanitarian Action (FATHUM) project. Trained in Anthropology and African Studies, Sara has a strong empirical orientation and has carried out "ethnographies of aid"—at the intersection of Science and Technology

Studies (STS), development theories, environmental anthropology and postcolonial studies—in which she broadly focused on how globally circulating ideas (such as climate change and notions of "modernity" and "development") travel, and what happens when they are translated by varying actors along the translation chain.

Freja C. Eriksen holds an MA in Journalism, Media and Globalization from Aarhus University and Universität Hamburg. Since concluding her thesis on sense-making of climate change in Greenland, she has become a climate and energy transition correspondent for *Clean Energy Wire* in Berlin. Before this, she worked as an editor at a Danish online news site covering the public sector. As a freelance journalist, she has researched the illegal international trade in electronic waste. She holds a BA in Rhetoric from the University of Copenhagen.

Thomas Friedrich received his doctorate in Social and Cultural Anthropology at Universität Hamburg within the framework of the interdisciplinary Cluster of Excellence "Climate System Analysis and Prediction". Previously, he was a fellow of the research group "The Cultural Constitution of Causal Cognition: Re-Integrating Anthropology into the Cognitive Sciences" at the Centre for Interdisciplinary Research in Bielefeld. He was a lecturer at the University of Cologne and the Ruhr-University of Bochum, and is now a researcher at the Institute for Social-Ecological Research in Frankfurt.

Imke Hoppe is a senior researcher at the Chair of Journalism and Communication Science at Universität Hamburg. She holds a PhD in Communication Science from TU Ilmenau. Her research interest lies in climate change and sustainability communication, with a special focus on digital media. Her research methods combine qualitative and quantitative methods of empirical communication research.

Shameem Mahmud works as a part-time lecturer in Media and Communication Research Methods and Journalism Cultures at the Institute of Journalism and Communication Studies, Universität Hamburg. He holds a PhD in Communication Studies from Universität Hamburg and specializes in public perceptions and communication of climate-change risks. Shameem also studied at the Universities of Dhaka, Bangladesh; Aarhus, Denmark; and Amsterdam, the Netherlands. His research focuses on environmental communication and journalism cultures. He previously worked at the University of Dhaka, and as a journalist in Bangladesh.

Friederike E. L. Otto is the Acting Director of the Environmental Change Institute (ECI) at the University of Oxford and an Associate Professor at the ECI Global Climate Science Programme. Her main research interests are extreme weather events and improving and developing methodologies to answer the question "whether, and to what extent, external climate drivers alter the likelihood of extreme weather to occur". She furthermore investigates the policy implications of this emerging scientific field. Friederike earned a Diploma in Physics from the University of Potsdam and a PhD in the Philosophy of Science from the Free University Berlin in 2012.

Simone Rödder is Assistant Professor of Sociology of Science and Principal Investigator in the Cluster of Excellence CLICCS at Universität Hamburg. She has an academic background in biology and sociology and is trained as a journalist. Her research explores communication across boundaries and focuses on science communication, science journalism, inter- and transdisciplinarity, and the science-policy interface. For recent publications see https://www.wiso.uni-hamburg.de/en/fachbereich-sowi/professuren/roedder/publikationen.html.

1. We are Climate Change
Climate Debates Between Transnational and Local Discourses

Michael Brüggemann and Simone Rödder

Local discourses around the world draw on multiple resources to make sense of a "travelling idea" such as climate change, including direct experiences of extreme weather, mediated reports, educational NGO activities, and pre-existing values and belief systems. There is no simple link between scientific literacy, climate-change awareness, and a sustainable lifestyle, but complex entanglements of transnational and local discourses and of scientific and other (religious, moral etc.) ways of making sense of climate change. As the case studies in this volume show, this entanglement of ways of sense-making results in both localizations of transnational discourses and the climatization of local discourses: aspects of the travelling idea of climate change are well-received, integrated, transformed, or rejected. Our comparison reveals a major factor that shapes the local appropriation of the concept of anthropogenic climate change: the fit of prior local interpretations, norms and practices with travelling ideas influences whether they are likely to be embraced or rejected.

https://doi.org/10.11647/OBP.0212.01

Silla [...] means the weather. Silla means also the human intelligence. So the weather and human intelligence are connected in our mythology, [...] the bigger intelligence that means the universe. So the universe, the weather of the globe, our weather [...] but also the human intelligence are connected.

Interview with Inuit representative on a United Nations climate summit, as quoted in Roosvall and Tegelberg (2018: 69).

The work of literary scholars, anthropologists, cultural historians, and critical theorists over the past several decades has yielded abundant evidence that "nature" is not nearly so natural as it seems. Instead, it is a profoundly human construction. This is not to say that the nonhuman world is somehow unreal or a mere figment of our imaginations— far from it. But the way we describe and understand that world is so entangled with our own values and assumptions that the two can never be fully separated. What we mean when we use the word "nature" says as much about ourselves as about the things we label with that word.

Cronon (1995: 25), as quoted in Jasanoff (2010: 245).

Echoing the first quote, it is a key proposition of the social sciences that representing nature entails representing humans (e.g. Jasanoff 2010; Jasanoff 2004; Luckmann 1970; Berger and Luckmann 1966). Anthropogenic global warming is a case in point: our physical environment, including the climate and the landscape, shapes our social realities; our social realities, in turn, influence our perceptions of our physical surroundings. These perceptions shape our practices and ways of living—which again affect the climate system.

"We are the climate", states ethnographer de Wit in Chapter 5, Tanzania. The phrase is echoed in different studies assembled in this book, including Maasailand in northern Tanzania, the capital city of Greenland, Nuuk, and the Philippine Island of Palawan.

There is, thus, a threefold connection between society and climate change. We are climate change, as our interpretations of climate change reflect who we are. We are climate change, as our lifestyle is based on the massive emission of greenhouse gases. Thirdly, and this is a major finding from the empirical studies presented in this book: the notion of a complex entanglement between society and nature, climate, and climate change is shared among many communities around the globe.

Given that climate change exists and will proceed regardless of what we think about it, why is it important to study how local communities make sense of climate change? As has long been argued in sociology (Luhmann 1989), the physical and biogeochemical processes described by terms such as *climate* and *climate change* have to be distinguished from the patterns of interpretation related to these processes that find resonance in society. In this sense, the concept of climate change is a social construction (Stehr and von Storch 1995). Interpretations of climate change, such as those that stress individual and collective efficacy (the belief that "we can make a difference"), may motivate people to change their lifestyles and, more importantly, mobilize political action, while feelings of fear and shock may overwhelm, paralyze actions or lead to risk denial (Feldman and Hart 2015; O'Neill and Nicholson-Cole 2009). Responsibility for action may be attributed to individuals or to political and economic decision-makers. Different causes of action may be advocated. All of these factors will ultimately influence how societies react to climate change. This is why it is not only of academic interest, but also of practical importance to study local discourses about climate change.

This book investigates sense-making as a process of social construction. Sociologists Berger and Luckmann (1966) have famously argued that society is composed of social constructions of reality in the myriad of day-to-day interactions in which we define, perform, and negotiate individual and collective selves. Social representations theory from social psychology likewise posits that there are patterns of meaning embedded in media content as well as in direct communication processes that shape social interactions (Höijer 2011, see also Chapter 2, Greenland). Combining this meaning-based approach with the idea of social differentiation, human geographer Hulme has described the scientific concept of climate change as "an idea that now travels well beyond its origins in the natural sciences. And as this idea meets new cultures on its travels and encounters the worlds of politics, economics, popular culture, commerce and religion—often through the interposing role of the media—climate change takes on new meanings and serves new purposes" (Hulme 2009: xxvi; see also Chapter 3, Philippines).

Important factors that influence both the individual and collective sense-making of climate change include the encounter of (1) transnational and local discourses and (2) scientific and other ways of sense-making. The two dimensions are empirically not distinct, yet it makes sense to distinguish them analytically and discuss them one by one.

(1) **Transnational and local discourses:** The climate debate is shaped by strong transnational (cutting across national borders) actors and institutions. The transnational character of climate research has been institutionalized in the set-up of and reports by the Intergovernmental Panel on Climate Change (IPCC), and is communicated based on a consensus policy (Hoppe and Rödder 2019). Another driver of transnational elements of climate debates is the United Nations Framework Convention on Climate Change and its annual climate summits (Conferences of the Parties, COPs). COPs have been found to be the most salient events in media coverage of climate change (Schäfer, Ivanova, and Schmidt 2013); they have become political media events reaching broad audiences (Brüggemann et al. 2017; Wessler et al. 2016).

Two observatories of media coverage of climate change at the Universities of Colorado (Media and Climate Change Observatory, MeCCO) and Hamburg (Online Media Monitor on Climate Change, OMM) confirm the relevance of COPs and transnational science events in drawing attention to climate change. They also show that climate change was heavily debated in the years 2007 to 2009, when the Nobel Peace Prize was awarded to the IPCC and Al Gore, and when the climate summit in Copenhagen attracted high hopes for transnational climate governance. The failure of Copenhagen led to a decade of public neglect of the issue, reflected in waning media attention to the topic (Boykoff et al. 2018; Brüggemann et al. 2018). The media debate on climate change has re-emerged in 2018 and 2019 after summers of heat and drought and the transnational climate strikes inspired by teenage activist Greta Thunberg (Mahl et al. 2020). The media debate is thus driven by transnational events and actors, including transnational NGO networks, as well as strategies of denial and downplaying the issue, fueled by the interests of carbon-dioxide-emitting industries (Oreskes 2017; Dunlap and McCright 2015).

While strong transnational discourses may lead us to expect similarities in local discourses about climate change, there are also good reasons to expect diversity. Local discourses are embedded in national political contexts that clearly also matter for the debate on climate change, as different content analyses have shown. National government positions tend to set the frames in national media outlets (Wessler et al. 2016; Grundmann and Scott 2014; Grundmann 2007). Ethnographic research has shown that local communities do not just mirror global climate discourses but develop their own interpretations of nature and/ or climate change (Stensrud and Eriksen 2019; Baer and Singer 2014; Crate and Nuttall 2009).

(2) Scientific and other ways of sense-making: The global governance approach of the climate regime resonates well with the prevailing scientific framing of the climate problem as global temperature rise (Aykut, Foyer, and Morena 2017; Aykut and Dahan 2015). Yet, scientific debates clash with the logics of other social worlds (Grundmann and Rödder 2019). The emergence of the concept of anthropogenic climate change has sparked entangled discourses in science, politics and journalistic media (Weingart, Engels, and Pansegrau 2000). National politics, as mentioned above, shapes national climate debates, as do the logics of journalism and political leanings of news outlets (Brüggemann and Engesser 2017; Boykoff 2011). Political ideologies overshadow scientific sense-making of climate change. Denying and downplaying the risks of anthropogenic climate change has become a defining feature of being Republican in the United States (Hoffman 2015; Dunlap and McCright 2008).

The impersonal and universal abstractions of science do not resonate well with local discourses. Jasanoff argues that the scientific-political representation of the climate problem as a global phenomenon is at odds with the sensations and memories through which local communities make sense of climatic change (2010: 237). The scientific meaning of climate change, as a decades-long increase in average global temperature, is not something that individuals can experience: "Global warming is not founded on everyday experience, has no immediate effects, and is not readily observable" (Ungar 1992: 489). Exposure to extreme weather or changes in vegetation or seasons can, of course, be experienced, yet it is an act of interpretation to link them to climate change (see also Rudiak-Gould 2013 on the controversy of whether climate change is visible). It has been shown that political concern about climate change has benefited from weather anomalies since the 1970s: "With a little help from Mother Nature [...] climate change research reached the agenda of top US policy-makers" in 1971-75 (Hart and Victor 1993: 665, as quoted in Grundmann and Stehr 2012: 120). Yet, extended phases of extreme summer heat (e.g., in 1988, 2012, and 2018/19) have only partly found resonance as "social scares" (Ungar 2014, 1992).

Communities do not necessarily draw a link between experiences of extreme weather and global warming, and indeed, connecting everyday experiences of weather phenomena to climate change has long been regarded as a misunderstanding of the scientific concept. Yet, as Jasanoff (2010) points out, this is an obvious way to help individuals make sense of the concept. Moving on from the misunderstanding-paradigm, a research field has recently emerged in climate sciences that explicitly aims to assess the connection between extreme weather phenomena and climate change, the science of event attribution (see Chapter 7, Attribution Science). To turn the link between extreme weather events and climate change into a new focus of climate research may be interpreted as a scientific response to the mismatch of scales between climate science and every-day experiences in both time and space.

Following Jasanoff, we assume that "the impersonal, apolitical and universal imaginary of climate change projected by science comes into conflict with the subjective, situated and normative imaginations of human actors engaging with nature" (2010: 233). This book addresses these tensions based on in-depth studies of how communities around the world make sense of climate change: How do local discourses relate to global discourses of science and politics? The volume's case studies extend beyond the well-researched Anglo-Saxon sphere and include both industrialized nations as well as perspectives from the Global South. Each case explores three dimensions of climate-change discourse.

The first dimension is *patterns of communication* related to climate change. Information about climate change often comes to us in mediated form, and the type of media influences the message that is conveyed (see, e.g., traditional media theories going back as far as McLuhan 1964). Therefore, each chapter analyzes who is communicating climate change messages—and using which media, including mass and social media, as well as interpersonal communications.

Second, we are interested in *patterns of interpretation* about climate change that emerge from the different flows of communication. Local recipients may engage in oppositional readings of media coverage (Hall et al. 1978): their sense-making might depart from the frames provided by elite sources that populate media coverage. Therefore, each chapter probes whether climate change is viewed in the study location as anthropogenic or not, and as a severe problem or not. Additional questions include: Who is held responsible for mitigating and/or adapting to climate change? Do people self-identify as victims of, or as contributors to, anthropogenic climate change? What solutions do they advocate?

The third dimension is the *entanglement of meanings* originating at the local or transnational level including how the scientific and other framings of climate change speak to each other.

We suggest grasping entanglements of meanings with the concepts of *localization of transnational climate discourses* and *climatization of local discourses*, respectively. The first term invokes the precedence of local patterns of interpreting concepts from afar. It has been used to describe processes or strategies that counter trends of homogenization through globalization and emphasizes the relevance of local factors of influence (see, e.g., Escobar 2001). The second term, climatization, stresses changes in local discourses induced by transnational climate-change discourses. It was developed in the context of studies on transnational climate policy making (Aykut, Foyer, and Morena 2017). Foyer and colleagues diagnose a "climatization" of the policy world whereby participants in various discourses present issues that were formerly unrelated to the climate problem through a "climatic lens", leading to the treatment of a variety of issues according to the dominant logics of the global climate regime (Foyer, Aykut, and Morena 2017).

Beyond this framework, the individual chapters draw on a variety of analytical perspectives. Each presents findings from an in-depth study (mostly PhD theses) conducted within different disciplinary traditions, including anthropology with its focus on detailed description (Geertz 1973, see Chapter 3, Philippines, and Chapter 5, Tanzania) and media and journalism studies, from both interpretive (see Chapter 2, Greenland, and Chapter 6, Bangladesh) and more standardized approaches (see Chapter 4, Germany). All share a solid grounding in fieldwork, and all prioritize local discourse, asking how communities around the world re-contextualize and reframe transnational discourses on climate change. As expected, all case studies demonstrate that local communities relate their everyday experiences of changing seasons and extreme weather events to climate change. Notably, the attribution of weather events to climate change has come full circle and refocused attention in climate science on event attribution. We therefore asked a climate scientist to introduce the field of event attribution science by presenting the current state of knowledge on how climate change affects extreme weather events and seasonal phenomena in the world regions discussed in this volume (see Chapter 7, Attribution Science). This final chapter can itself be read as a piece of scientific discourse—written with a broad audience in mind—that speaks to the other chapters' explorations of local communities' sense-making.

In the following, we summarize each study, explain its approach and case-specific findings and eventually discuss some common patterns that emerge from the case studies.

Climate Change as a Vehicle for Hopes and Fears in Greenland

While the Arctic regularly serves as a poster child for dangerous anthropogenic climate-change impacts in media reports, its citizen's perspectives on both the impacts and their media representation have so far received much less attention (but see Nilsson and Christensen 2019; Nuttall 2009). Chapter 2, written by communication scholar Freja Eriksen, helps fill this gap by studying how Greenlanders make sense of climate-change impacts through both media exposure and personal experiences. Her analysis is based on social representations theory and on five focus group interviews involving fifteen inhabitants of Nuuk, the capital of Greenland.

Eriksen finds that while iconic images of melting icebergs portray Arctic citizens as climate change victims, her interviewees do not self-identify as victims. The local discourses she describes diversify the global imaginary of Greenland "melting away", even including the positive image of a future capital "Costa del Nuuk". The notion of "Costa del Nuuk" is an example of naming and anchoring the unclear future of a warming Greenland with well-known concepts such as summer holidays on the Spanish Mediterranean coast. These humorous replies confirm previous findings by Nuttall (2009), whereby political aspirations for self-governance and independence are linked to a warming climate that opens up chances for future development (while Denmark, the former colonizer, might become flooded).

Eriksen's findings also echo Nuttall's (2009) argument that in Greenland, the climate is widely understood to be permanently changing and intrinsically unstable. Similar to perceptions among the Maasai (see Chapter 5, Tanzania), unreliable seasons and a changing climate are thus not believed to be extraordinary, new, or necessarily manmade.

Notably, for the inhabitants of Greenland, who we might expect to have direct experiences with climate change, the topic appears heavily mediated by news coverage. As one high school student puts it: "We live in Greenland but we don't see icebergs break off every day". While the interviewees criticize too stereotypical media descriptions of climate-change impacts, media coverage remains a major reference point. And while some interviewees oppose the dominant journalistic story of harmful climate change with Greenland as one of the most heavily affected countries, Eriksen points out that the alternative narrative put forward—politically and economically beneficial climate change—may also be taken from the media.

The chapter highlights that both media and local discourses cannot be treated as homogeneous entities. Eriksen found that an individual's professional background influenced whether they stressed the potential economic benefits or environmental risks of climate change. Also, groups of older people were more likely to doubt that humans have triggered current climate change and to believe that the media exaggerate climate change. Younger interviewees asserted that the media underestimate the true dangers of man-made global warming. Yet both age groups shared a distrust and criticism of media coverage, albeit for different reasons. Some inhabitants called for more local Greenlandic perspectives on climate change in the media.

The chapter clearly illustrates both the clash between transnational and local discourses and the conflict between the scientific view and self-assessments of climate-change impacts. While science warns that the Arctic is more heavily affected by climate change and is warming faster than most other world regions, and transnational media stereotypes propagate iconic images of polar bears and melting ice, local discourses emphasize that the climate has always been changing, that warmer temperatures might actually be better for Greenland and that Greenlanders have been able to adapt to these changes throughout history.

Recontextualizing a Travelling Idea: Climate Change on the Philippine Island of Palawan

Chapter 3, written by social and cultural anthropologist Thomas Friedrich, studies the case of the Philippine island of Palawan to shed light on how climate change is perceived on an island where environmental protection is popular. Employing multi-method ethnography, the chapter explores how the idea of climate change is locally reframed based on personal experiences, and embedded in pre-existing ways of knowing nature and cultural practices including an already strong environmentalism.

Conceptually, the chapter draws on Hulme's interpretation of climate change as a "travelling idea" (2009), and argues that the notion of climate change has evolved predominantly in a top-down direction, from global IPCC knowledge via networks of media and politics to local people with diverse cultural backgrounds and epistemologies who try to make sense of it. Friedrich combines this notion with a Science and Technology Studies-informed distinction between knowledge and meaning, reiterating Jasanoff's point that scientific climate knowledge arises from impersonal observation whereas cultural meanings emerge from embedded experience in specific environmental, social and cultural contexts (Jasanoff 2010).

Because the island of Palawan has been exposed to severe environmental destruction, the government has pursued a vigorous environmental discourse and a determined environmental policy since the 1990s, which has also contributed to a consensual view of the perception of climate change.

The chapter examines educational theatre performances to demonstrate how the production and dissemination of a local notion of climate change takes place. The communicator of knowledge about climate change that the ethnographer examined in this case is a local NGO. In all their plays, natural disasters were linked to morally wrong environmental behavior such as cutting mangroves for charcoal production, an illegal but widespread practice on the island. The author thus argues that activities like the theatre "contribute to maintaining the popularity of tree planting as a means of environmental and climate protection". Recommending already widespread activities helps to secure the popularity of environmental protection measures, as opposed to discourses that brand popular practices, such as eating meat or driving cars, as environmentally harmful.

As for the question of entanglements between local and global discourses, the chapter shows that there is no linear relationship between climate knowledge and environmentally friendly behavior. On Palawan, climate change is perceived as one natural hazard among many, and "the discourse on climate change may have served more as post-hoc justification than original motivation for past and present behavior". It is thus neither knowledge of, nor belief in, anthropogenic climate change that makes a behavioral difference. Rather, the global discourse reinforces pre-existing beliefs, values, and practices. On Palawan, "it strengthens people's traditionally strong environmentalism and validates their strong rejection of, for example, cutting trees or burning garbage". In such a situation, the climate may be protected 'accidentally' while communities pursue other goals.

Climate Change Comes Closer for Locals in Northern Germany

In Chapter 4, a media and communication studies author team (Imke Hoppe, Fenja De Silva-Schmidt, Michael Brüggemann, and Dorothee Arlt) reports findings from a broader research project at Hamburg University that explored audience reception of the COP 21 climate summit in Paris. The chapter compares how the Paris summit has been perceived and interpreted in an urban (Hamburg) and a rural setting (Otterndorf), both located in Northern Germany. The small town of Otterndorf is situated at the end of the Elbe estuary on the North coast, and the metropolis of Hamburg is located upstream along the Elbe River. The authors explore how space, both as a physical and a social concept, influences interpretations of climate change, with a focus on the role media reception plays in the process. The chapter draws on data from focus group discussions and media diaries, and uses an online panel survey from the wider project to contextualize the qualitative analysis presented in the chapter.

The results show that both media use and interpretations of climate change are fairly similar in both settings. Respondents in both settings criticized local media coverage of the summit. Apparently, local newspapers failed to explain the complex matter of transnational climate governance and make it locally relevant. Locals were aware of the summit but claimed that the media did not explain what the international negotiations were about.

Spatial differences mattered as residents of Hamburg and Otterndorf differed in the extent to which they were personally concerned about potential climate-change impacts. Even though sea level rise is a tangible consequence of climate change for residents of Hamburg as well, they hold strong beliefs that it will not affect them personally. Historical floods (devastating parts of Hamburg in 1962) are no longer viewed as a threat. People in Otterndorf, with the rising tides of the North Sea behind the dikes, felt more personally concerned about climate change and worried that coastal protection was insufficient. Local discourses found in Hamburg thus provide evidence of the belief that Hamburg (e.g., through the harbor) contributes to climate change, but is not really affected. The local discourse in Otterndorf, taking place among the inhabitants and apparently being neglected by local media, was more about climate change as a threat to some of their villages.

An unexpected finding was the role of local roots and connection: long-term inhabitants drew stronger links between climate change, climate policy making, and their local community. The study thus concludes that the longer an individual lives in a place and the more connected he or she feels to it, the more relevant spatial factors become for her or his experience of climate change. While the science of anthropogenic climate change is widely accepted, the global policy discourse remains remote to local communities in Germany.

Resistance to the Idea of Anthropogenic Climate Change among the Maasai

Chapter 5 travels south into Maasailand. It draws on fourteen months of ethnographic work conducted by anthropologist Sara de Wit for a PhD dissertation at the University of Cologne. She studies the ways in which climate-change discourse is translated, communicated and received in a rural village in Northern Tanzania, exploring how villagers who have no experience with Western life and whose culture is shaped by religion translate the story of climate change.

Conceptually, the chapter draws on cultural theories of risk perception (e.g., Douglas 1992) that emphasize that our models of nature and risk are not value-neutral scientific descriptions but are influenced by moral and political considerations. The chapter reconstructs the translations of the transnational discourse about climate change driven by the mass media (represented by the local radio station), the (Christian) church and NGOs. It finds that the ways in which climate science is translated do not sit well with the Maasai's religion and culture. The absence of awareness of and talk about climate change is not primarily rooted in a lack of knowledge, argues de Wit, but constitutes an "attempt to remain faithful to one's own set of norms, values, beliefs".

The first reason for the Maasai to oppose the scientific narrative about climate change is that it is perceived as an attack on their religion. In the local language, the same word (*Eng'ai*) denotes God, sky (or heaven), and rain. Drought and rain are the domain of God. Now NGOs—in an educational movie described in the chapter—are telling the Maasai that it is climate change that brings about drought and that humans cause climate change. Hence, it is of no use to pray for rain. Scientists are viewed as secular prophets propagating their apocalyptic tale. Also, discussing the future meets resistance among the Maasai, who "refuted any attempt to probe future climate scenarios or ideas of the future in any sense. Questions related to futurology were always cast away with laughter, followed by 'we cannot know', or 'only God knows'".

The second reason why the message of anthropogenic climate change meets opposition is that the local community views changing weather and unreliable seasons as normal conditions of life. The idea of stability as the "natural" state of nature seems strange (see also Chapter 2, Greenland). A third reason is that the Maasai currently have to deal with a deep cultural change following a switch from a nomadic to a semi-nomadic lifestyle: "Perhaps the weather has changed, but we have changed, too". The deterioration of climatic conditions is furthermore interpreted as a consequence of the decline of culture and morals.

De Wit points out that in this context it is "not fruitful to disentangle climatic and societal changes"—a marked contrast to the scientific attribution approach as outlined in Chapter 7, which focuses on isolating the contribution of climate change to extreme weather events. She concludes that the story of climate change is not doomed to be rejected locally, if it respects the "inclusive ontology in which society, morals and nature are interwoven—a way of living that ceases to make sense when purged of *Eng'ai*".

Climate Change and Other Troubles along Bangladesh's Coastline

In Chapter 6, Bangladesh-born communication scholar Shameem Mahmud focuses on the country's coastal region to explore how communities at the forefront of climate-change risks make sense of the concept. Thus, the chapter deals with yet another group that is often dubbed "climate victims" in Western discourses.

The chapter studies the community's major information sources on climate change and how it understands climate change in the context of constant exposure to regional geo-hazards such as tropical cyclones, floods, salinity in the water and soil, storms, and coastal erosion. The conceptual part of the chapter is based on the climate change communication and risk perception literature. Methodologically, the study draws mainly on interviews with thirty-eight citizens of a coastal district in Bangladesh, located close to the world's largest mangrove forest. Almost half of the participants were illiterate.

The study finds that interviewees obtained climate change information from a number of sources, including mass media (radio and television), NGO advocacy programmes, and local opinion leaders. This acquired knowledge focuses on topics such as salinization and sea level rise as local impacts of climate change. It is interesting to note that, just as in Greenland (see Chapter 2, Greenland), the local communities in this climate-change-affected country still learn about the concept through the media. An interesting aspect of this chapter is the role of NGOs in framing the climate issue for the coastal community. There are 250 active NGOs in the region, and attributions to climate change are often induced by NGO officials who conduct awareness programmes for farmers and fishermen. "I have heard this might be because of climate change" was a standard narrative among community members in response to the interview questions.

Mahmud identifies two major patterns of sense-making of climate change within the community. First, a "regional geo-hazard pattern", which contextualizes risks from climate change with well-known local geo-hazards, namely "climate change as increased and unprecedented storms", "climate change as drowning in the sea", and "climate change means increased salinity". The second pattern is "weather and seasonal variance". It describes how the interviewees attribute their personal experiences of changes in local weather and seasons to climate change. The communities' direct exposure to extreme weather events and experiences of seasonal change and other vagaries of the weather clearly influence how they make sense of mediated climate change information.

In their explanation of phenomena, such as increased salinity and rising tidal surges, the interviewees stated that the phenomena were "not new", but attributed the increase of the problem to the hitherto abstract concept of climate change—a phenomenon that social representations theory (see Chapter 2, Greenland, and Chapter 3, Philippines) describes as "anchoring" a new concept in familiar ones. The interviewees were also aware of local causes other than climate change: e.g. shrimp aquaculture as a source of increased salinization. Rather than attributing any ecological problem to the global issue of climate change, and in contrast to some of what they have heard from NGOs, the community members claimed agency, and responsibility: "We should not always blame others for the problem".

The study reveals complex entanglements between local and global discourses, including the question of responsibility for a problem like salinization. The chapter highlights the importance of analyzing NGOs' communication strategies, tools and content, as well as their role in creating awareness and motivating behavioral changes among rural communities. In conclusion, local "place identity" formed through experiences of regional geo-hazards, in combination with media information and NGO communication, make climate change a salient risk in this coastal community of Bangladesh.

Attributing Extreme Weather Events to Climate Change: The Perspective of Science

In Chapter 7, physicist and climate modeler Friederike Otto describes the kinds of changes that are already being experienced locally and those that we are likely to see in the future. As one of the pioneering researchers in the emerging field of event attribution, she shows how research has recently advanced in exploring the link between extreme weather events and climate change. The chapter also exemplifies how science produces knowledge that is subsequently considered as authoritative in society. The process starts with a "landmark paper" that proposes a new methodology, which is followed by peer scrutiny and results in published claims of relevant knowledge.

The chapter translates the question of links between climate change and extreme weather into the scientific language of changing probabilities: climate change has increased the likelihood of particular types of events by a certain percentage. It also emphasizes that attributable changes in these probabilities are associated with a number of uncertainties, and that the robustness of various studies varies strongly depending on the types of events and geographic region investigated.

Nonetheless, there are some basic facts that climate scientists now treat as known for certain, including the fact that the current global warming trend is anthropogenic. This warming causes rising sea levels and increases the risk of some extreme weather events. This link is more evident for heat waves and large-scale rainfall, while at this time droughts and other events can only be linked to climate change in specific regions and seasons.

While Otto clearly stresses advances in this field, she also highlights limitations such as the complexity of weather phenomena like droughts and a lack of data, especially in the Global South. There are also natural variabilities in the climate system that influence how often extreme events occur. Furthermore, anthropogenic climate change interferes with other human impacts such as changing land use patterns and how communities manage rivers and water systems (see Chapter 6, Bangladesh). Otto explains that the damage caused by extreme events is understood as an outcome not only of the event itself, but local vulnerability and exposure are major risk factors in addition to the meteorological hazards. The methodological rationale of the scientific approach is to disentangle these factors in order to accurately attribute given effects to particular causes. This approach to disentangling different causes stands in marked contrast to how different factors (anthropogenic, natural) and considerations (scientific, religious, moral, and economic) are entangled in local debates about climate change.

This chapter draws on data gathered from around the world and provides a perspective that is shaped by the logic of scientific research. Although it holds insights on specific regions, its perspective is not that of a local community but of a distinct professional way of sense-making. Still, it illustrates how the media and public debates influence climate science, and thus provides evidence of what has been conceptualized as the medialization of science (Rödder et al. 2012; Weingart 1998). Individual perceptions and journalistic claims about links between extreme weather events and climate change have long preceded climate scientists' ability to draw this link. Public claims about possible links may have provided the impetus for the emergence of the science of event attribution—which has only very recently matured to make claims about the role of anthropogenic climate change in specific weather events.

Comparing Local Discourses

Local discourses on climate change are, in many ways, unique. Describing and understanding the specifics and details of each case is an endeavor that goes beyond the scope of book chapters and deserves full-length dissertations (De Wit 2017; Friedrich 2017; Mahmud 2016). Having summarized the cases with at least some of their specific results, we now briefly discuss them comparatively, looking at similarities and differences. We follow our framework of *patterns of communication, patterns of interpretation,* and *patterns of entanglement of transnational and local discourses.*

Patterns of Communication: How Local Communities Use and Evaluate Sources of Information on Climate Change

Local communities typically distinguish between "us" (voices from my village, family, friends) and "them/others" (voices from afar, such as journalistic media, NGOs, and scientists) in the use and assessment of sources on climate change. Some sources, like local media outlets or local NGOs, may be situated in between these two categories, depending on how strongly they are rooted locally. "Other" sources are used to learn about the world beyond the local community ("I have seen it on TV", "It was on the news"), yet these sources are not trusted in the same way as those socially close to home. When asked whom they would rather believe about climate change, students from Greenland responded their local "elders" or "my grandmother" (see Chapter 2, Greenland). In more traditional societies, being socially close seems to overlap with spatial proximity, since families and communities tend to be located in the same village or in a limited area. In highly mediatized Western societies, particularly in urban contexts, being spatially and socially close (in the sense of belonging to the same community) may or may not overlap. Yet, the distinction between familiar sources that are part of the community and voices from afar remains relevant in both settings.

Media coverage, in particular, is criticized, mainly for *not* being able to understand local concerns related to climate change. Being a *detached* observer (and therefore keeping distance) is, in fact, a core self-understanding of journalism (Hanitzsch 2011; Deuze 2005). Climate change as a travelling idea that comes from afar is typically mediated by journalistic media, as well as by the NGOs' educational and activist programmes. The localization of the transnational discourse may sometimes fail, and remain remote and irrelevant to local publics.

Explicit criticism of news reporting was salient in our Western case studies where interviewees claimed that the media (including local media) provided a stereotypical image of them as helpless victims of climate change (Greenland), or failed to make the issue of global warming locally relevant (Northern Germany). In line with the well-researched phenomenon of a "hostile media" perception (Gunther and Schmitt 2004), some respondents assume that the journalistic media serve some kind of hidden agenda (such as protecting the profits of ocean liners in Hamburg, see Chapter 4, Germany). We also see plural (negative) evaluations of the coverage: while some inhabitants of Nuuk criticized the media for being too alarmist, others opposed them for neglecting the risks associated with climate change (see Chapter 2, Greenland).

Several chapters highlight the role of NGOs in bringing ideas of climate change to local communities. As the chapters describe, it is mainly through educational programmes and events that local or western NGOs try to create climate-change awareness and motivate behavioral change among local communities. The chapters about cases from the Global South do not relate any instances of explicit criticism of either media content or NGO activities.

Differences between northern and southern societies occur, possibly, due to the higher degree of mediatization in the northern societies, where all areas of life are shaped by digital media use (see, e.g., Couldry and Hepp 2016). Thus, the students from Greenland emphasize that they do not see melting icebergs in everyday life, but know about them from media coverage.

The direct information activities of NGOs were not mentioned in our western case studies. It may be—and this needs to be further studied— that the NGO's local information activities in the Global South serve as a functional equivalent to mediated communication in western societies.

Patterns of Interpretation: How Local Communities Make Sense of Climate Change

Many informants in the communities under study share the feeling of being affected by climate change. This is expressed as a change from a more stable and better past to an unpredictable present and a potentially threatening future. Most explicitly articulated among the Maasai, but also elsewhere, these changes are attributed to a moral decline in society. A complex nature-culture connection via morality and religion is a commonplace finding in the anthropological literature, typically idealizing the past in order to criticize the present (Schnegg 2019).

While the local communities analyzed in the case studies feel affected by climate change, they do *not* see themselves as victims. Rather, they claim agency and feel competent to fight climate change, for example by praying or planting trees. And they feel able to adapt. This finding is in line with other studies which have found that local communities self-identify as "designers of the environment", and is in stark contrast with the transnational discourse that casts them as "helpless victims of climate change processes" (Jurt et al. 2015). Yet, there is little denial of the anthropogenic nature of climate change (some Maasai and some respondents in Greenland are exceptions that we will return to below), and there is an overall willingness to address the problem.

Another cross-cutting finding is that local discourses are not homogeneous: they are divided by age group, social position of the interviewee, and degree of rootedness in the local community. Older people seem to be more inclined to resist new narratives such as climate change, and we find evidence of this in Greenland as well as among the Maasai. Studies have shown this for other countries, even though there are national differences in the relative importance of each of the factors influencing perceptions of climate change (Poortinga et al. 2019; Capstick et al. 2014).

Entanglement of Local and Transnational Discourses

The chapters report instances of both the localization of transnational discourses and the climatization of local discourses. Transnational climate discourses travel to communities some of which do not read or write and its meanings are transformed along the way, resulting in vastly diverging localizations around the world. The chapters on Greenland and Germany emphasize that local subgroups differ in how they make sense of climate change. The Greenland case study shows how younger audiences are more connected to transnational discourses through journalistic and social media. The German case study emphasizes differences between a village and a city, as well as between inhabitants with deeper or shallower roots in the local community. The localization of discourses in this case affects inhabitants with deeper roots in the community to a greater extent.

At the same time, we see instances of "climatized" local practices. Chapter 3 finds that the global climate discourse fosters the already popular practice of collective tree planting and thus "recycles" this practice "in climatic terms" (Foyer and Kervran 2017). The Maasai, however, mostly reject the idea of anthropogenic climate change, which conflicts with God's responsibility for rain, the weather, and the climate. This is notably similar to the views of some fundamentalist Christian communities in the United States (Shao 2017; Smith and Leiserowitz 2013). In Bangladesh and the Philippines, by contrast, the idea of climate change falls on culturally responsive grounds. Whether a scientific idea like climate change is embraced is thus linked to the respective community's prior beliefs about nature, humankind and the universe. Comparing the Maasai, the cases in Bangladesh and the Philippines hint at an important driver of the climatization of discourses: the *fit of prior local interpretations, norms and practices with travelling ideas* influences whether they are likely to be embraced or rejected.

We find an overall primacy of local concerns, which may be explained by drawing an analogy of local communities' sense-making with individual psychological mechanisms. Confirmation bias and avoidance of cognitive dissonance explain how individuals preserve their world views; similar processes may also apply to how local communities preserve their own discourses. New information is selectively used to confirm pre-existing cultural ideas, such as tree planting in the Philippines or urging people not to cut down the mangrove forest in Bangladesh.

All chapters confirm noticeable differences between the epistemologies employed by science and those of local communities, drawing, e.g., on religion and morality. The major feature of how people draw on different resources to make sense of climate change is indeed relating the idea of climate change from global scientific and policy discourses to personal or collective experiences of weather events or seasonal changes. It is interesting to note that climate science, with the field of event attribution, has adapted to this common-sense logic of relating the abstract (climate change) to the concrete experience (extreme weather event). Departing from emphasizing the difference between weather and climate, scientists now also focus on trying to explain how weather events and climate change relate (see Chapter 7, Attribution Science).

When discussing the changes they witness, local communities often express a sense of "shared responsibility" between carbon dioxide emissions from afar and ecologically deleterious local practices, such as shrimp farming (in Bangladesh), cutting trees (in the Philippines), or personal consumption patterns (in Germany). People on the coast of Bangladesh do not forget the local sources of some of their ecological problems (salinization), and do not attribute these problems solely to climate change. For this reason, the tendency of some NGOs to attribute a given local problem to global climate change at times clashes with the local communities' attributions of local problem causes. The attribution of a local problem to climate change and, often, the promotion of technical fixes as prime solutions, calls into question the communities' own ways of knowing, as well as claiming agency and attributing responsibilities. Thus, while climate protest movements may have the best of intentions, attributing concrete local problems solely to climate change depoliticizes practices on the local level and may help local authorities deviate attention and deflect responsibility away from their own failures and mismanagement.

Many ideas and practices in local discourses, like planting trees to avoid climate change, are not fundamentally at odds with scientific scenarios of mitigating climate change, but an important aspect is often lost in translation: the scale of the issue. Politicians and scientists talk about global challenges, while individuals integrate their conclusions from this global discourse into their local day-to-day activities. While some interviewees in the Philippines and Bangladesh assume that the problem can be tackled if they plant some trees, some current climate scenarios would require transforming areas the size of entire countries into industrial tree-growing plantations. There is again a mismatch between the scale of scientific forecasts on the infrastructure changes necessary to combat climate change and the actions people can take within the constraints of their everyday lives.

Finally, we find that individuals' reception of the idea of climate change does not depend on their education. The majority of respondents from Bangladesh and the Philippines may be less educated and may have more pressing concerns than climate change, but they do not doubt its existence. Partly, they engage in climate protection 'accidentally', by continuing the cultural practice of tree planting that preceded their awareness of climate change (see Chapter 3, Philippines). These cases once more confirm the finding that there is no simple link between scientific literacy, climate-change awareness, and a sustainable lifestyle (Grémillet 2008).

Outlook

Local discourses around the world draw on different resources to make sense of climate change, including mediated reports, experiences of extreme weather, and pre-existing values and belief systems. This results in both localizations of transnational debates and the climatization of local discourses and practices.

One may think about this research not only for the purpose of understanding local discourses and their entanglement with scientific and political climate debates, but it may inspire communication practice, as well. When communicating climate change, one should be ready to accept that people have more urgent concerns to deal with in their everyday lives—but sometimes climate change may be related to these more obtrusive concerns. Identifying links between the global problem and local concerns is useful for developing strategies for engaging people with climate change. Planting trees has been practiced for good ecological reasons before worrying about climate change, yet it also benefits climate protection. When trying to solve ecological problems on the coast of Bangladesh, an academic discussion of whether salinization is due to climate change or due to shrimp farming is beside the point, as *both* of these (and probably other) factors contribute to the problem. In local discourses, as De Wit's chapter shows, seemingly contradicting approaches to interpreting the world co-exist. Schnegg (2019), in a study on how pastoralists in Namibia explain rain patterns as opposed to how meteorologists explain the same phenomenon, finds that individuals do not have to rigorously decide between the two competing knowledge systems; they can relate to both, depending on their respective social context. They can "switch between ways of knowing, and thus between worlds". Whether people effectively do so is also likely to depend on whether one knowledge system (whether it is science or religion, for instance) claims precedence in all domains of life. This insight has consequences for how climate change should *not* be communicated: as a modern substitute for religious beliefs. This might be more easily achieved by a local communicator who can complement, and possibly, integrate, a scientific perspective with the belief systems (religious or not) of a local community.

We encourage our readers to draw their own conclusions on how to communicate climate change at home, wherever that may be. The following case studies show that knowledge about the locally salient discourses around nature and climate change is a precondition for understanding how to communicate climate change. They also, hopefully, inspire a more interdisciplinary and transdisciplinary dialogue on how local communities make sense of climate change.

References

Aykut, Stefan C., and Amy Dahan. 2015. *Gouverner le climat?: Vingt ans de négociations internationales*, 2nd edn (Paris: Presses de Sciences)

Aykut, Stefan C., Jean Foyer, and Edouard Morena (eds). 2017. *Globalising the Climate: COP21 and the Climatisation of Global Debates*, Routledge Advances in Climate Change Research (London: Routledge), https://doi.org/10.4324/9781315560595

Baer, Hans A., and Merrill Singer. 2014. *The Anthropology of Climate Change: An Integrated Critical Perspective* (London: Routledge), https://doi.org/10.4324/9781315818702

Berger, Peter L., and Thomas Luckmann. 1966. *The Social Construction of Reality* (New York: Anchor Books)

Boykoff, Max, Midori Aoyagi, Anne G. Ballantyne, Andrew Benham, and Patrick Chandler. 2018. World Newspaper Coverage of Climate Change or Global Warming: 2004-2020. Media and Climate Change Observatory Data Sets. Cooperative Institute for Research in Environmental Sciences, University of Colorado, https://doi.org/10.25810/4c3b-b819

Boykoff, Max. 2011. *Who Speaks for the Climate?* (Cambridge, UK: Cambridge University Press), https://doi.org/10.1017/cbo9780511978586

Brüggemann, Michael, F. De Silva-Schmidt, I. Hoppe, Dorothee Arlt, and Josephine B. Schmitt. 2017. "The Appeasement Effect of a United Nations Climate Summit on the German Public", *Nature Climate Change*, 7.11: 783–87, https://doi.org/10.1038/nclimate3409

Brüggemann, Michael, Irene Neverla, Imke Hoppe, and Stephanie Walter. 2018. „Klimawandel in den Medien", in *Hamburger Klimabericht—Wissen über Klima, Klimawandel und Auswirkungen in Hamburg und Norddeutschland*, 107, ed. by Hans von Storch, Insa Meinke and Martin Claußen (Berlin, Heidelberg: Springer Berlin Heidelberg), pp. 243–54, https://doi.org/10.1007/978-3-662-55379-4_12

Brüggemann, Michael, and Sven Engesser. 2017. "Beyond False Balance: How Interpretive Journalism Shapes Media Coverage of Climate Change", *Global Environmental Change*, 42: 58–67, https://doi.org/10.1016/j.gloenvcha.2016.11.004

Capstick, Stuart, Lorraine Whitmarsh, Wouter Poortinga, Nick Pidgeon, and Paul Upham. 2014. "International Trends in Public Perceptions of Climate Change over the Past Quarter Century", *WIREs Climate Change*, 6: 35–61, https://doi.org/10.1002/wcc.321

Couldry, Nick, and Andreas Hepp. 2016. *The Mediated Construction of Reality* (Cambridge, Malden: Polity)

Crate, Susan A., and Mark Nuttall (eds). 2009. *Anthropology and Climate Change: From Encounters to Actions* (Walnut Creek, CA: Left Coast Press)

Cronon, William (ed.). 1995. *Uncommon Ground: Rethinking the Human Place in Nature* (New York: W.W. Norton and Co)

De Wit, Sara. 2017. "Love in the Times of Climate Change: How the Idea of Adaptation to Climate Change Travels to Northern Tanzania" (PhD thesis, University of Cologne)

Deuze, Mark. 2005. "What is Journalism? Professional Identity and Ideology of Journalists Reconsidered', *Journalism*, 6.4: 442–64, https://doi.org/10.1177/1464884905056815

Douglas, Mary (ed.). 1992. *Risk and Blame: Essays in Cultural Theory* (London, New York: Routledge)

Dunlap, Riley E., and Aaron M. McCright, 2015. "Challenging Climate Change: The Denial Countermovement", in *Climate Change and Society: Sociological Perspectives*, ed. by Riley E. Dunlap and Robert J. Brulle (New York, NY: Oxford University Press), 300–32

Dunlap, Riley E., and Aaron M. McCright. 2008. "A Widening Gap: Republican and Democratic Views on Climate Change", *Environment: Science & Policy for Sustainable Development*, 50.5: 26–35, https://doi.org/10.3200/ENVT.50.5.26-35

Escobar, Arturo. 2001. "Culture Sits in Places: Reflections on Globalism and Subaltern Strategies of Localization", *Political Geography*, 20.2: 139–74, https://doi.org/10.1016/S0962-6298(00)00064-0

Feldman, Lauren, and P. Sol Hart. 2015. "Using Political Efficacy Messages to Increase Climate Activism", *Science Communication*, 38.1: 99–127

Foyer, Jean, Stefan C. Aykut, and Edouard Morena. 2017. "Introduction: COP21 and the 'Climatisation' of Global Debates", in *Globalising the Climate: COP21 and the Climatisation of Global Debates*, Routledge Advances in Climate Change Research, ed. by Stefan C. Aykut, Jean Foyer, and Edouard Morena (London: Routledge), pp. 1–17 https://doi.org/10.4324/9781315560595-1

Foyer, Jean, and David D. Kervran. 2017. "Objectifying Traditional Knowledge, Re-enchanting the Struggle against Climate Change", in *Globalising the Climate: COP21 and the Climatisation of Global Debates*, Routledge Advances in Climate Change Research, ed. by Stefan C. Aykut, Jean Foyer, and Edouard Morena, (London: Routledge), pp. 153–72, https://doi.org/10.4324/9781315560595-9

Friedrich, Thomas. 2017. *Die Lokalisierung des Klimawandels auf den Philippinen: Rezeption, Reproduktion und Kommunikation des Klimawandeldiskurses auf Palawan* (Wiesbaden: Springer VS), https://doi.org/10.1007/978-3-658-18232-8

Geertz, Clifford. 1973. *The Interpretation of Cultures: Selected Essays* (New York: Basic Books)

Grémillet, David. 2008. "Paradox of Flying to Meetings to Protect the Environment", *Nature*, 455.7217: 1175, https://doi.org/10.1038/4551175a

Grundmann, Reiner. 2007. "Climate Change and Knowledge Politics", *Environmental Politics*, 16.3: 414–32, https://doi.org/10.1080/09644010701251656

Grundmann, Reiner, and Simone Rödder. 2019. "Sociological Perspectives on Earth System Modeling", *Journal of Advances in Modeling Earth Systems*, 11.12: 3878–92, http://dx.doi.org/10.1029/2019MS001687

Grundmann, Reiner, and Mike Scott. 2014. "Disputed Climate Science in the Media: Do Countries Matter?", *Public Understanding of Science*, 23.2: 220–35, https://doi.org/10.1177/0963662512467732

Grundmann, Reiner, and Nico Stehr. 2012. *The Power of Scientific Knowledge: From Research to Public Policy* (Cambridge, UK: Cambridge University Press), https://doi.org/10.1017/cbo9781139137003

Gunther, Albert C., and Kathleen Schmitt. 2004. "Mapping Boundaries of the Hostile Media Effect", *Journal of Communication*, 54.1: 55–70, https://doi.org/10.1111/j.1460-2466.2004.tb02613.x

Hall, Stuart, Chas Critcher, Tony Jefferson, John Clarke, and Brian Roberts. 1978. *Policing the Crisis* (London: Macmillan)

Hanitzsch, Thomas. 2011. "Populist Disseminators, Detached Watchdogs, Critical Change Agents and Opportunist Facilitators: Professional Milieus, the Journalistic Field and Autonomy in 18 Countries", *International Communication Gazette*, 73.6: 477–94, https://doi.org/10.1177%2F1748048511412279

Hart, David M., and David G. Victor. 1993. "Scientific Elites and the Making of US Policy for Climate Change Research, 1957–74", *Social Studies of Science*, 23.4: 643–80, https://doi.org/10.1177/030631293023004002

Hoffman, Andrew J. 2015. *How Culture Shapes the Climate Change Debate* (Stanford, California: Stanford Briefs, an imprint of Stanford University Press)

Höijer, Birgitta. 2011. "Social Representations Theory: A New Theory for Media Research", *Nordicom Review*, 32.2: 3–16, https://doi.org/10.1515/nor-2017-0109

Hoppe, Imke, and Simone Rödder. 2019. "Speaking with One Voice for Climate Science—Climate Researchers' Opinion on the Consensus Policy of the IPCC", *Journal of Science Communication*, 18.3: A4, https://doi.org/10.22323/2.18030204

Hulme, Mike. 2009. *Why We Disagree about Climate Change: Understanding Controversy, Inaction and Opportunity* (Cambridge, UK: Cambridge University Press)

Jasanoff, Sheila. 2004. "Heaven and Earth: The Politics of Environmental Images", in *Earthly Politics: Local and Global in Environmental Governance, Politics, Science, and the Environment*, ed. by Sheila Jasanoff and Marybeth L. Martello (Cambridge, MA: MIT Press), pp. 31–52

——. 2010. "A New Climate for Society", *Theory, Culture & Society*, 27.2–3: 233–53, https://doi.org/10.1177%2F0263276409361497

Jurt, Christine, M. D. Burga, L. Vicuña, et al. 2015. "Local Perceptions in Climate Change Debates: Insights from Case Studies in the Alps and the Andes", *Climatic Change*, 133.3: 511–23, https://doi.org/10.1007/s10584-015-1529-5

Luckmann, Thomas. 1970. "On the Boundaries of the Social World", in *Phenomenology and Social Reality*, ed. by Alfred Schutz and Maurice A. Natanson (Dordrecht: Springer), pp. 73–100, https://doi.org/10.1007/978-94-011-7523-4_5

Luhmann, Niklas. 1989. *Ecological Communication* (Chicago: University of Chicago Press)

Mahl, Daniela, Michael Brüggemann, Lars Guenther, Fenja De Silva-Schmidt. 2020. "Public Opinion at a Tipping Point: Germany's Path to Engaging with Climate Protection", *Down to Earth*, 16 April, https://www.fdr.uni-hamburg.de/record/851#.X5V1z0eSmUk

Mahmud, Shameem. 2016. "Public Perception and Communication of Climate-change risks in the Coastal Region of Bangladesh: A Grounded Theory Study" (PhD thesis, Hamburg University)

McLuhan, Marshall. 1964. *Understanding Media: The Extensions of Man*, 5th edn (Cambridge, MA: MIT Press)

Nilsson, Annika, and Miyase Christensen. 2019. *Arctic Geopolitics, Media and Power*, Routledge Geopolitics Series (London, New York, NY: Routledge)

Nuttall, Mark. 2009. "Living in a World of Movement: Human Resilience to Environmental Instability in Greenland", in *Anthropology and Climate Change: From Encounters to Actions*, ed. by Susan A. Crate and Mark Nuttall (Walnut Creek, CA: Left Coast Press), pp. 292–310

O'Neill, Saffron, and Sophie Nicholson-Cole. 2009. "'Fear Won't Do It': Promoting Positive Engagement with Climate Change Through Visual and Iconic Representations", *Science Communication*, 30.3: 355–79, https://doi.org/10.1177%2F1075547008329201

Supran, Geoffrey, and Naomi Oreskes. 2017. "Assessing ExxonMobil's Climate Change Communications (1977–2014)", *Environmental Research Letters*, 12.8: 84019, https://doi.org/10.1088/1748-9326/aa815f

Poortinga, Wouter, Lorraine Whitmarsh, Linda Steg, Gisela Böhm, and Stephen Fisher. 2019. "Climate Change Perceptions and their Individual-Level Determinants: A Cross-European Analysis", *Global Environmental Change*, 55: 25–35, https://doi.org/10.1016/j.gloenvcha.2019.01.007

Rödder, Simone, Martina Franzen, and Peter Weingart (eds). 2012. *The Sciences' Media Connection: Public Communication and its Repercussions*, Sociology of the Sciences Yearbook 28 (Heidelberg: Springer), https://doi.org/10.1007/978-94-007-2085-5

Roosvall, Anna, and Matthew Tegelberg. 2018. *Media and Transnational Climate Justice: Indigenous Activism and Climate Politics* (New York: Peter Lang), https://doi.org/10.3726/b13285

Rudiak-Gould, Peter. 2013. *Climate Change and Tradition in a Small Island State: The Rising Tide* (New York, London: Routledge), https://doi.org/10.4324/9780203427422

Schäfer, Mike S., Ana Ivanova, and Andreas Schmidt. 2013. "What Drives Media Attention for Climate Change? Explaining Issue Attention in Australian, German and Indian Print Media from 1996 to 2010", *International Communication Gazette*, 76.2: 152–76

Schnegg, Michael. 2019. "The Life of Winds: Knowing the Namibian Weather from Someplace and from Noplace", *American Anthropologist*, 26.2: 183, https://doi.org/10.1111/aman.13274

Shao, Wanyun. 2017. "Weather, Climate, Politics, or God? Determinants of American Public Opinions toward Global Warming", *Environmental Politics*, 26.1: 71–96, https://doi.org/10.1080/09644016.2016.1223190

Smith, Nichols, and Anthony Leiserowitz. 2013. "American Evangelicals and Global WarmingW, *Global Environmental Change*, 23.5: 1009–17, http://dx.doi.org/10.1016/j.gloenvcha.2013.04.001

Stehr, Nico, and Hans von Storch. 1995. "The Social Construct of Climate and Climate Change", *Climate Research*, 5: 99–105, https://doi.org/10.3354/cr005099

Stensrud, Astrid B., and Thomas H. Eriksen (eds). 2019. *Climate, Capitalism and Communities: An Anthropology of Environmental Overheating* (London: PlutoPress), https://doi.org/10.2307/j.ctvjnrw0q_

Ungar, Sheldon. 1992. "The Rise and (Relative) Decline of Global Warming as a Social Problem", *The Sociological Quarterly*, 33.4: 483–501, https://doi.org/10.1111/j.1533-8525.1992.tb00139.x

——. 2014. "Media Context and Reporting Opportunities on Climate Change: 2012 versus 1988", *Environmental Communication*, 8.2: 233–48, https://doi.org/ 10.1080/17524032.2014.907193

Weingart, Peter. 1998. "Science and the Media", *Research Policy*, 27.8: 869–79, https://doi.org/10.1016/s0048-7333(98)00096-1

Weingart, Peter, Anita Engels, and Petra Pansegrau. 2000. "Risks of Communication: Discourses on Climate Change in Science, Politics, and the Mass Media", *Public Understanding of Science*, 9.3: 261–83, https://doi.org/10.1088/0963-6625/9/3/304

Wessler, Hartmut, Antal Wozniak, Lutz Hofer, and Julia Lück. 2016. "Global Multimodal News Frames on Climate Change", *The International Journal of Press/Politics*, 21.4: 423–45, https://doi.org/10.1177/1940161216661848

2. The Case of "Costa del Nuuk"

Greenlanders Make Sense of Global Climate Change[1]

Freja C. Eriksen

This chapter investigates how fifteen inhabitants of the Greenlandic capital, Nuuk, make sense of climate change and its impacts through media exposure and personal experiences. While Greenland's melting ice sheet has long served as a backdrop to the global climate debate, local public views of climate change have largely been overlooked. This study finds that, although the media is an important source of information about climate change for the inhabitants of Nuuk, their sense-making of the phenomenon is saturated by personal experiences. Alarmist media representations, for instance, are continuously challenged by references to personal experiences of positive local impacts of climate change. The chapter identifies six distinctions underlying the inhabitants' sense-making of climate change—*natural/unnatural, certainty/uncertainty, self/ other, local/global, positive/negative,* and *environment/economy.*

---

1 This chapter is based on the author's Master's thesis, submitted in 2016 at Universität Hamburg.

 https://doi.org/10.11647/OBP.0212.02

A "Poster Child" for Climate Change

"In a couple of years, you can come up here to visit 'Costa del Nuuk'"[laughs].

Johannes

This chapter presents a case study of how fifteen inhabitants of Greenland's capital, Nuuk, make sense of climate change and its impacts through media exposure as well as personal experiences. It opens with a quote from one of the study's participants, Johannes. He is fifty-nine years old, works at an airline in Nuuk, and cannot help but laugh at the thought of climate change transforming his hometown into a sunny holiday destination. This study contributes his and fourteen other interviewees' sense-making of climate change to the literature on public views of climate change. According to the Intergovernmental Panel on Climate Change (IPCC), "human influence on the climate system is clear" (IPCC 2014: 2). However, there are several gaps in our understanding of how climate change is experienced, understood, and made sense of in its many contexts around the world. The case of Nuuk illuminates only one in a daunting range of cases yet to be studied (Schäfer and Schlichting 2014). It is, however, particularly pertinent due to the observable and rapid melting of Greenland's ice sheet and the country's status as a "poster child for Arctic climate change" (Holm 2010: 145).

Greenland (Kalaallit Nunaat) and its melting ice sheet have for several years served as a backdrop to the international climate debate for politicians, environmental campaigners, journalists and researchers (Bjørst 2011). The annual shrinkage of the ice sheet, which covers 80% of the country's geography, has "increased four-fold from 1995–2000" (Arctic Monitoring and Assessment Programme 2012: 3). In the foreign media, Greenland regularly headlines stories, such as "Greenland is melting away" (Davenport et al. 2015) and "Climate change: Greenland loses a trillion tons of ice in four years as melting rate triples" (Harvey 2016). Studies of British tabloid and broadsheet newspapers have documented how images associated with Greenland, of "melting ice" and "polar bears", have come to dominate the coverage of climate-change impacts (Smith and Joffe 2009). The status of these iconic images is underlined by studies that show how laypeople's first associations of climate change in the USA, England, and Sweden were references to "melting glaciers and polar ice", "polar regions melting", "melting icebergs" and "rising temperatures" (Wibeck 2014b; Smith and Joffe 2013; Leiserowitz 2006). Through climate change science, international news coverage, and global institutions such as the IPCC, a "climate change crisis discourse" has developed, in which the Arctic nature, and, to a lesser extent, its citizens, play a leading role (Farbotko and Lazrus 2012; Bravo 2009; Martello 2004). While in this discourse "climate vulnerable populations are being positioned as victims, but also as evidence of the climate crisis", their own voices and perspectives are at risk of drowning (Farbotko and Lazrus 2012: 382). But although such narratives are prominent in global climate-change discourses, they may not necessarily represent Arctic citizens' views (Bravo 2009). Studying how the people of Greenland make sense of climate change and its impacts is therefore both relevant and compelling.

Greenland has arguably been on a path to "greater economic and political independence" (Government of Greenland) since the end of Danish colonization in 1953. Yet, its economy still greatly depends on an annual block grant from Denmark of about 470 million euro (Government of Greenland). The desire for economic and political independence is a powerful undercurrent in the country's public debate on climate change (Bjørst 2011; Nuttall 2009). In April 2016, Greenland opted for a territorial reservation to the Paris Agreement, as negotiated in December 2015, because the agreement did not take the country's economic development considerations fully into account (Government of Greenland). The disappearance of ice in Greenland has made valuable resources, such as oil, gas, mining and hydrocarbon development, both at sea and on land, more accessible (Ackrén and Jakobsen 2015; Nuttall 2009). Due to these possibilities for economic development, Nuttall (2009: 295) concludes that "Greenland is literally warming to the idea of less snow and ice". Greenland represents a particularly interesting case to study: in the global climate governance regime, it must balance its role as the voice of an Indigenous people, while campaigning for its right to produce emissions and thus contribute to climate change (Bjørst 2008).

I share Farbotko and Lazrus' (2012: 382) view of climate change as both a "discursive and material phenomenon", and find that Nuuk's inhabitants are affected by climate change in both a physical and a social sense. Nuuk's residents represent a "climate-exposed population" (Farbotko and Lazrus 2012: 382) whose voices and views are an important addition to the current literature; global climate-change narratives must take local discourses and individual experiences of climate-change impacts into account. According to Farbotko and Lazrus, the "[c]limate is changing, but its meanings are contingent on place and history and cannot be imposed from above without risk of disjunctures and injustice" (2012: 383). This paper hence proceeds on the assumption that Greenland's political discourses on climate change *and* its citizens' sense-making of the issue should be added to current research on local, contextual understandings of climate change around the world. The study's main research question thus is *how do inhabitants of Nuuk make sense of climate change and its impacts through media exposure and personal experiences?*

I investigate this question using (1) data gathered in five focus group interviews involving a total of fifteen participants and (2) a survey of their media use, both in general and related to climate change. The chapter proceeds by introducing the study's relevant concepts and theoretical framework of social representations. I then review previous studies of audience reception of climate change coverage and outline the study's research design and methodology. The bulk of the chapter is dedicated to presenting and discussing the study's findings. The main conclusion is that, although the media is an important source of information about climate change for Nuuk's inhabitants, their sense-making of the phenomenon is heavily influenced by personal experiences. For instance, they continuously challenge alarmist media representations by referring to personal experiences of positive local impacts of climate change. Finally, the chapter identifies six distinctions underlying residents' sense-making of climate change—*natural/ unnatural, certainty/uncertainty, self/other, local/global, positive/negative,* and *environment/economy*.

Conceptual Considerations and Literature Review

"Climate-Exposed Populations"

While research on public views of climate change has generally focused on North America and Europe, there have been some recent studies on Asia, Africa, and Latin America (Schäfer and Schlichting 2014; Wibeck 2014a). Schlichting and Schäfer's 2014 review of 133 publications does not mention any studies of Greenland, although they stress the relevance of analyzing the countries "most affected by" climate change (2014: 146). Most climate change-related studies of Greenland thus far have been conducted by "physical scientists, often with a focus on the ice sheet" (Holm 2010: 145; Hall et al. 2006; Lüthcke et al. 2006; Zwally et al. 2002). Studies involving Greenlanders' attitudes towards climate change are fewer, although growing in number (Holm 2010).

As climate-change impacts—and efforts to mitigate those impacts— have entered the global agenda, an interest in Indigenous peoples and their role in anthropogenic climate change has emerged (Martello 2008). Studies have gradually investigated Indigenous peoples' perceptions of climate change and how their "traditional ecological knowledge" can inform our understanding of physical climatic changes (Roosvall and Tegelberg 2012). A group of anthropological researchers has touched on local climate-change discourses in Greenland (Tejsner 2013; Bjorst 2011; Holm 2010; Nuttall 2009, 2008; Leduc 2007). Using multi-sited ethnography, Bjørst (2011) identifies several recurring discourse elements framing climate change in relation to Greenland and the Arctic. Studying the perceptions of her informants in the Disco Bay area, she finds a common "mistrust in science" and a related questioning of the phenomenon of climate change in general, which has left room for more local theories of climate change (Bjørst 2011). She and Nuttall (2009: 297–98) maintain that climate is often understood to be continuously changing and intrinsically unstable, related to the Greenlandic word for climate (*sila*). As such, climate change is not necessarily perceived as extraordinary or manmade (see Chapter 5, Tanzania). This outlook leads to a situated form of "normalization of the threat" (Bjørst 2011: 242).

Another key observation made by both Nuttall (2009) and Bjørst (2011: 243, author's translation) is a preoccupation in Greenland with "possibilities as well as problems". Nuttall (2008: 47) argues that this apprehension is perhaps more prevalent in Greenland than anywhere else in the circumpolar North, due to its aspirations for self-governance and independence, which are made more likely by the prospect of a warmer climate opening up new industries. Bjørst (2011: 246, author's translation) similarly identifies a tendency to put "development, national political and economic independence before anything else". These are important findings, because it can be assumed that these tendencies resonate in the local sense-making of climate change. To my knowledge, this study is the first to examine Greenlandic discourses of climate change from a communication and media studies perspective.

Following the finalization of the present study, a nationally representative survey of Greenlandic perspectives on climate change has, however, been published (Minor et al. 2019). It provides evidence that more than nine out of ten Greenlanders think climate change is happening, although only about half think it is mostly caused by human activities and about a third believes it is caused by "natural changes in the environment". Residents in the area of the survey covering Nuuk (West Sermersooq) were slightly more certain that climate change is happening and is caused mostly by human activities. Two thirds of the residents of West Sermersooq said they had experienced the effects of climate change (compared to 76% in the general Greenlandic population), and about 60% said they heard about climate change in the media at least once a week or at least once a month. 45% heard people they know talk about climate change at least once a week or at least once a month. The survey also reveals a clear tendency of respondents to see climate change as mainly a bad thing. In West Sermersooq, about half saw climate change as either "bad" or "very bad", while most others saw it as "neutral". A greater part of the respondents also saw climate change as more likely to harm than benefit the people of Greenland. A majority of respondents both in West Sermersooq and Greenland in general saw "protecting the environment even if it costs jobs" as more important than "economic growth, even if it leads to environmental problems", favoring policies of re-entering the Paris Agreement, regulating greenhouse gas emissions from industry and investing in alternative energy sources. These findings are pertinent as they form a more recent and representative insight into perspectives on climate change in Greenland and Nuuk where environmental concerns apparently outweigh economic interests.

Social Representations Theory

The present study contributes to the constructionist literature that has used social representations theory to research "public views" of climate change (Wibeck 2014b; Smith and Joffe 2013; Olausson 2011; Whitmarsh, Lorenzoni, and O'Neill 2011; Cabecinhas, Lázaro, and Carvalho 2008; Lorenzoni and Pidgeon 2006; Moscovici 1984a). As these studies have demonstrated, social representations theory is particularly suitable to researching how laypeople or specific groups within society make sense of climate change. At its core, the theory, originated by Moscovici (1984b), endeavors to explain how people familiarize themselves with hitherto unknown objects or change. Moscovici's (1984a) theory presupposes that "concepts and images" of scientific origin permeate society, where they evolve into collective and "mundane understandings" (Olausson 2011: 953). The social representations approach views the popular sense-making of, for example, scientific knowledge, as "a valid knowledge system in its own right", as opposed to a distortion or inaccurate representation of scientific knowledge (Moloney et al. 2014: 1; Wibeck 2014b: 206). Following this approach, the present study does not seek to verify whether citizens' knowledge of climate change is *scientifically accurate*, but to understand how the scientific concept of climate change is *publicly made sense of* (Smith and Joffe 2013). I use the phrase "make sense" to operationalize the process of creating meaning, as employed in Wibeck's (2014b), Olausson's (2011), Moloney et al.'s (2014), and Marková's (2003) adaptations of Moscovici's theory.

According to Moscovici (1984a: 953), social representations "fill our minds and our conversations, our mass media, popular books, and political discourses". Therefore, mass media have only increased the importance of social representations by multiplying the changes ideas undergo as they are subject to media simplifications and sensationalizing (Moloney at al. 2014; Moscovici 1984b). The media is an important "discursive site" where social representations are "(re)produced" (Olausson 2011). Analyzing interviewees' social representations reveals a glimpse of "iconic images" that circulate in the "socio-cultural context" (Smith and Joffe 2013).

This study relies on four central notions to operationalize the theory of social representations. According to Moscovici (1984b), the process of making the unfamiliar familiar happens through two central mechanisms—"objectifying" and "anchoring". *Objectifying* refers to a mechanism of turning "something abstract into something almost concrete" (Moscovici 1984b: 29); it involves attributing a real-life form to the abstract (Moscovici 2000). For example, climate change is typically objectified in media representations as a storm, heatwave, or flood (Höijer 2010). *Anchoring* is a process in which people place new ideas in relation to categories or paradigms of their existing representations (Moscovici 2000). In other words, new ideas are placed in a "familiar context" in order to make sense of them (Moscovici 1984b). The process of anchoring takes place through "naming" and "classifying", also known as "drawing distinctions" (Moscovici 2000: 42; Olausson 2011: 285). Through the mechanism of *naming*, abstract phenomena are given a "label" or "familiar name" (Moscovici 2000: 42). According to Moscovici (2000: 42), a phenomenon that does not have a familiar name or category is perceived as "alien" and "threatening". Thus, for example, laypeople refer to climate change as "weather" in some studies. In this way, the "abstract and intangible risk" it poses "acquires familiar and comprehensible characteristics" (Olausson 2011: 289). *Drawing distinctions* involves placing intangible phenomena in the context of "well-known opposites", in a form of hierarchy or classification (Olausson 2011: 285; Moscovici 2000: 43). Smith and Joffe (2013: 19) label these distinctions "themata", and define them as "mutually interdependent oppositions or dialogical antinomies". Common distinctions, such as *natural/unnatural, certainty/uncertainty* and *self/other*, reveal the "latent content, or latent drivers of public thinking" around climate change among interviewees (Smith and Joffe 2013: 16).

Making Sense of Climate Change

The present study employs social representations theory to investigate how the inhabitants of Nuuk make sense of climate change. Most similar to this study's conceptual and methodological approach are Olausson's (2011) and Wibeck's (2014b) case studies in Sweden. Both studies explore laypeople's social representations of climate change through focus group interviews.

Olausson (2011: 289) finds that respondents discuss climate change's "existing consequences" with great conviction and use "everyday experiences of the weather" as a way to anchor their conceptualizations of the phenomenon, often familiarizing it through the label "weather". She concludes that "the respondents' own experiences are of utmost importance", which is especially relevant for the climate-exposed sample of Nuuk's inhabitants (Olausson 2011). Wibeck (2014b: 209, 211) similarly reports participants referencing their "own experiences of changes in weather", but stresses the finding that they saw climate change mostly as a "global issue" with severe, yet distant consequences. She finds that focus groups continuously negotiate the certainty and severity of climate change and our ability to mitigate it (2014b: 212–14). Somewhat in contrast, Olausson finds that a general "belief" in anthropogenic climate change seems to be commonplace among her interviewees (2011: 286).

Smith and Joffe (2013: 21–22) find "melting ice" to be their London participants' most common first association with climate change, followed by references to "weather" and "pollution". These associations, they conclude, echo "how newspapers visually represent the threat" in British media. They identify three "themata" that interviewees use to structure their sense-making in a "non- or unconscious" way (2013: 19, 22). First, they categorize their British interviewees into those who are certain and those who are uncertain about anthropogenic climate change. While respondents generally express certainty that global warming is real, many are still uncertain about its causes, and blame this on "contradictory media coverage" (Smith and Joffe 2013: 27). Second, they identify a thema they call "natural/unnatural", which highlights the way respondents observe "strange and bizarre weather" in relation to their expectations of how nature should behave, thus also drawing on personal experience of changes in weather patterns (Smith and Joffe 2013: 25-26). Third, they identify a thema labelled "self/other" (Smith and Joffe 2013: 23-25), and find respondents place responsibility for climate change mostly on the "other", portraying the USA, China, and India, for example, as the main perpetrators of global warming. The impacts of climate change are also largely distanced from the "self", and serious consequences are thought to affect others more (Smith and

Joffe 2013: 25). Meanwhile, notions of the "self" are related to solutions to climate change, often associated with recycling, which ambiguously provide either a sense of "satisfaction" or "helplessness and frustration" (Smith and Joffe 2013: 25).

Audience Reception of Climate Change Coverage

The scientific construct of global climate change is not directly visible, and therefore requires "media transmission", even to Greenlanders (Taddicken 2013). There is limited understanding of the media's role in shaping the public's "knowledge [of] and attitudes" towards climate change (Taddicken 2013: 39; Cabecinhas, Lázaro, and Carvalho 2008: 172). Taddicken argues that until now, "communication scholars have paid much greater attention to media content than to audiences and media users" (2013: 40). In many current studies within the field, the media is assumed to play a key part in framing people's understanding of climate change, but these assumptions are "rarely verified with reference to empirical studies on the relationship between media output and audience reception" (Olausson 2011: 282). Hence, there is a need for more research on the "complex reception process" that takes place as people make sense of a phenomenon such as climate change—a process in which the media "constitutes only one of several meaning-making resources" (Olausson 2011: 282). The current study therefore adopts a similar approach to that of Olausson (2011: 282–83), viewing the media as a central element in the analysis, while preventing a focus on the content of media reporting on climate change from overshadowing the audience's complex reception process.

Olausson concludes that the media serves as her Swedish focus group participants' primary source of information about climate change, although this information is "negotiated and remolded in conversations and discussions with other people" (2011: 294). She refers to the media as the "agenda-setter", but argues it is important not to underestimate how people negotiate and embed their own experiences into media discourses and vice versa when making sense of climate change (Olausson 2011). Olausson further identifies three criticisms of general media representations noted by her Swedish interviewees: a fatigue with "emotionally charged media messages", an awareness of their "commercial conditions", and the media coverage's "lack of continuity and integration of various sorts of news" (2011: 292–93). Ryghaug, Sorensen, and Naess (2011: 784) similarly conduct focus group interviews with members of the Norwegian public, and find that all groups referred to the media (specifically newspapers (print or digital) and "television and radio") as their primary source of climate change information. Yet their interviewees, like those in the Olausson (2011) study, believe that the media exaggerates stories "to sell more newspapers" (Ryghaug, Sorensen, and Naess 2011: 790).

Previous studies have portrayed the mass media—first, TV; second, newspapers; and, to a growing extent, the internet—as the main sources of information on climate change, which are more influential than "people's interpersonal communication" (Schäfer 2015: 853). These conclusions, however, are based on a review of surveys from the USA, the UK, Germany, and Australia (Schäfer 2015). In Southern England, Whitmarsh (2008) found mass media to be the main source of information on climate change among respondents, with TV as the primary source, followed by newspapers and radio, while Stamm, Clark, and Eblacas (2000) found newspaper and TV to be the most used information sources in a segment of US citizens. In Portugal, Cabecinhas, Lázaro, and Carvalho (2008: 174) found that the media (television news, followed by newspapers and "televised films and documentaries") was their survey participants' main source of information on climate change. Respondents viewed the news media and "people they know" as somewhat credible on this issue, while they mistrusted "government, local authorities and corporations". Taddicken (2013: 39) investigated the "impact of mass media and internet use" on German internet users' knowledge and attitudes towards climate change and found that only regular use of television, but not of print media, positively affected knowledge and attitudes. She concluded that visual cues are powerful tools for communicating information about climate change. Arlt, Hoppe, and Wolling (2011: 52, 57) found that watching public news programmes in Germany predicted greater awareness, while reading weekly print media "had a slightly negative effect on problem awareness", indicating that media use does "not always have a motivating, awareness-heightening effect". A survey of the German public by Brüggemann et al. (2017) showed limited effects of media coverage of annual UN climate summits on knowledge and attitudes, and no effect on the respondents' intention to act.

Research Design and Methodology

Research Design

Based on the conceptual considerations and findings from previous studies, this study explores how inhabitants make sense of climate change and its impacts. It is an open inquiry into how interviewees represent climate change and its impacts at both the local and global levels. It explores interviewees' associations with both the scientific phenomenon of climate change and its impacts (e.g., socio-economic), including how these associations are negotiated among focus group members, how important interviewees believe future climate change is, how they demarcate it, etc.

The study furthermore explores the origins of interviewees' representations of climate change by investigating the relevance of personal experiences versus media exposure in shaping their sense-making of climate change and its impacts. "Personal experiences" is defined as interpersonal communication, as well as personal climatic and societal observations; in sum, anything that inhabitants refer to in making sense of climate change and its impacts that is not directly related to mediated communication. "Media" refers to any type of mass media content consumed by focus group participants, including printed newspapers, internet sources, radio, TV, social media, Netflix documentaries, etc.

Finally, the study explores *how* the interviewees relate media representations to their personal experiences, and how they negotiate any contradictions between information gained from the two sources. The aim is to determine how inhabitants interpret media representations, such as how they relate to memories of specific examples of media coverage or general evaluations of discourses on climate change and Greenland's role in these.

The research questions were investigated in August 2016 through (1) five focus group interviews with fifteen participants and (2) a survey of eleven of these participants' media use, both general and related to climate change. Unfortunately, the survey responses of FG 5 could not be retrieved. The case study method is useful for investigating a phenomenon that cannot be separated from "important contextual conditions"—in this instance, local discourses of climate change in Greenland's capital city. This object of study is intrinsically linked to its "real-world context", as we are interested in how discourses of a global, scientific phenomenon are shaped in a local environment (Yin 2014). The case study approach allows the use of different sources of evidence (Yin 2014). The present study employs a mixed-methods approach that draws on data from qualitative focus group interviews and a survey of the same participants' media use (David and Sutton 2011). The focus group interviews give insights into the interviewees' "everyday experiences, meanings and language", while the self-administered survey explores the participants' individual media use more directly and in depth than would be possible during a focus group (David and Sutton 2011).

The Case Study: The City of Nuuk

Located in southwest Greenland, Nuuk is the world's northernmost capital (Gill 2015). The city is home to one-quarter of Greenland's 55,847 inhabitants (Rasmussen 2016: Schultz-Lorentzen and Rasmussen 2012; Statistikbanken). Nuuk has a variety of inhabitants employed in "government administrative work, education, health care [...] hunting, fishing, fish and shrimp processing" (Encyclopaedia Britannica 2017). The city encompasses a cross-section of professions, ages and geographic backgrounds. This study thus complements previous findings from Greenland that have mostly studied Disco Bay, located close to the receding glaciers of Ilulissat Icefjord. Greenlandic is the population's first language, while 12% of Greenlanders are bilingual Danish speakers (Rischel 2016). Half of Greenland's Danish inhabitants live in Nuuk (Statistikbanken).

Focus Group Interviews

The study's primary method was five semi-structured focus group interviews with two to four participants in each group (Table 2.1) (David and Sutton 2011). According to Moscovici, it is precisely "material from samples of conversations [which] gives access to the social representations" (2000: 62). Previous studies of sense-making of climate change have employed this technique (Wibeck 2014b; Olausson 2011; Ryghaug, Sorensen, and Naess 2011). A total of seven men and eight women were interviewed. Workplaces and institutions were contacted by email and phone asking for three to four people willing to take part in the study. The targeted workplaces and institutions were chosen according to the methodological principle of maximum variance, taking into consideration that people with different professions, which are presumably affected by climate change to varying degrees, might express different perspectives.

Two administrative employees at a fish factory in Nuuk (FG 1) were sampled as fishery is Greenland's largest industry and about half of the country's fishing fleet is based in Nuuk (Schultz-Lorentzen and Rasmussen, 2012). Three airline employees were included in the sample (FG 4) as tourism comprises a growing industry in Greenland, while air traffic by its nature contributes to the country's CO_2 emissions. Both industries were assumed to be somewhat affected by climate change, and thus the views of their employees were deemed of interest. A bank clerk and two women working in kindergartens, albeit in different schools, were targeted (FG 2) in order to include insights of inhabitants whose work would not be directly affected by climate change. As more than 60% of Nuuk's inhabitants work in public and private service, these also constitute important segments of the population (Schultz-Lorentzen and Rasmussen, 2012). Two Danish government officials were chosen in order to add insights from inhabitants who had obtained higher education and to provide a perspective of someone dealing, perhaps indirectly, with climate change as a political matter. A Greenlandic tourist guide was included to get a perspective from someone in contact with the tourists in Greenland, some of them specifically to experience climate change (FG 3). Four high school students were included in the sample (FG 5) in order to get the perspectives of a group of younger inhabitants of the city. While climate change has been a known phenomenon in Greenland for much of their lives, the younger generation are also an interesting group as their media use could vary from that of the generation before them. Since employees and students signed up to participate in the interview, a small self-selection bias (e.g., interest in climate change) must be expected and accounted for in the analysis.

Table 2.1 Focus group participants

Group no.	Participants	Participants (no. and gender)	Name and age	Nationalities	Interview date
FG 1	Administrative employees in fishing industry	2 (2 men)	Frederik (53), Bjarne (56)	Danish, Greenlandic,	9/8/2016
FG 2	Employees in public and private service (banking/education)	3 (3 women)	Jonna (55), Nivi (49), Karen (64)	Greenlandic, Greenlandic, Greenlandic	18/8/2016
FG 3	Public officials and tourist guide	3 (3 men)	Malthe (23), Gorm (59), Ulrik (34)	Greenlandic, Danish, Danish	20/8/2016
FG 4	Employees at different levels of airline	3 (1 woman, 2 men)	Sofie (38), Johannes (59), Pavia (55)	Greenlandic, Greenlandic, Greenlandic	17/8/2016
FG 5	High school students	4 (3 women, 1 man)	Ivalu (n/i*), Tanja (n/i*), Hansine (n/i*), David (n/i*)	n/i*	17/8/2016

Table adapted from Wibeck (2014b: 208).
*no information.

As the study's research question refers to "inhabitants of Nuuk", Danes are included in this sample. Nuuk is home to half of all of the country's inhabitants born outside of Greenland and has a high population of Danes. Many of the Danish inhabitants in the study have, however, resided in Greenland for several decades. Interviews were conducted in Danish, which is the second language for most Greenlanders (Rischel 2016) and the first language of the author. A translator was not necessary. Most interviewees were fluent in Danish, while a few were helped by other interview participants to remember forgotten words. Participants' names, workplaces with few employees and place names mentioned during the interviews were anonymized during transcription.

All interviews were conducted according to a semi-structured interview guide. They took between 45 and 80 minutes, well within the recommended time frame of 45 to 90 minutes (David and Sutton 2011). All five group interviews were slightly different, as the interviewer attempted to follow the flow of the participants' conversation as well as follow up on interesting themes raised. The interviews were audio taped, transcribed and subsequently analyzed by qualitative content analysis in order to map the "patterns within [the] qualitative data" (David and Sutton 2011). Deductive codes were based on the analytical elements of social representations theory (*objectification, anchoring, naming,* and *distinctions*), and key themes of interest according to the research questions. Subsequently, a "descriptive" coding, an "interpretative" and a final coding—including inductive sub-codes—were undertaken (David and Sutton 2011; King and Horrocks 2010).

Survey on Media Use

Little research exists on media use in Greenland. One of the few previous studies has shown that TV and radio are the population's most important media sources, in particular the Greenlandic Broadcasting Corporation's (KNR) programmes (Ravn-Højgaard et al. 2018). People living in Nuuk are, however, found to watch less TV and listen half as much to the radio than the general population. Meanwhile, the capital's inhabitants use the internet more (68%) than the rest of the Greenlandic population (43%), and social media are used more in Nuuk (56% on a daily basis as compared to 43 % in the rest of the country). Also, younger Greenlanders are found to use the internet more, and TV and KNR's services less. 35% of the population primarily receive news through online media, a trend which is growing.

To gain a deeper understanding of the current study's interviewees' differentiated media use in relation to general news and climate change specific news, all participants were asked to fill out a quantitative survey (David and Sutton 2011). The survey sought information on participants' stated media use, sources of information on climate change, and their evaluation of the sources' trustworthiness—information that would not have been fully uncovered in a focus group interview. By combining and comparing qualitative interview and quantitative survey data, the study carefully develops its understanding of how participants make sense of climate change through media and personal experience (David and Sutton 2011; Greene, Caracelli and Graham 1989).

The first half of the survey consisted of three parts, asking the interviewees about: (1) their media use related to news about Greenlandic and international affairs, (2) their media use related to information about climate change, and (3) which sources they would trust most to answer questions about climate change. The second half of the survey consisted of demographic questions about the interviewees' gender, age, nationality, years lived in Nuuk, geographic heritage, civil status, education, employment situation, and type of employment. The survey was developed using elements of Jensen and Helles' (2015) and Metag, Füchslin, and Schäfer's (2015) surveys. The survey was given to the study's participants upon finishing the focus group interview for practical reasons, and to avoid priming the focus groups in any direction.

Findings

The following analysis is structured around the study's major research questions. Within each section, relevant themes are analyzed by exploring participants' ways of objectifying and anchoring climate change through naming and distinctions. Since social representations are said to become accessible through "material from samples of conversations", we seek to discover them in the interactions between focus group participants (Moscovici 2000: 62). The analysis pays close attention to both the groups' collective and separate individuals' patterns of sense-making. According to Moscovici (2000: 63), social representations are "revealed especially in times of crisis and upheaval, when a group or its images are undergoing a change". Hence, special attention is paid to situations in which interviewees disagree, discuss, or contest with each other (Wibeck 2014b). At the same time, it is relevant if absolute agreement or a complete absence of tension is observed. This would suggest that social representations within an area are widely agreed upon to the extent that they are no longer consciously reflected on (Moscovici 2000).

Making Sense of Climate Change and its Impacts

The first part of the study's findings explores the inhabitants of Nuuk's immediate associations with and definitions of climate change. How do they relate the concept to the context of Greenland and Nuuk? In short, how do they make sense of the phenomenon?

Natural and Unnatural Local Weather Phenomena: Objectifications of Climate Change

The first interview started with the question, "What is the first thing that comes to your mind when someone mentions climate change?" The first images evoked in interviewees' minds by such an open question can reveal which prior social representations of climate change interviewees draw on while they are still relatively uninfluenced by the researcher's agenda (Wibeck 2014b). While the participants' responses varied, many immediately related climate change to the local context of Greenland and changes therein that participants had seen or heard about. An illustration of this pattern is found in four high school students' answer to the initial question:

> Tanja: Global warming
> David: That's, that's the same thing actually. [laughs]
> Ivalu: Uhm, I think hunters, don't know why, always hunters. Poor hunters.
> David: Mmm, I think about shorter winter.
> Tanja: Warmer summer.
> Ivalu: Also, stormy weather. We have that often now, compared to before, like real storms.

Hansine: Really bad, bad weather in the winter.
Tanja: And the ice sheet, when it melts.
Ivalu: It's alright if it melts. [all laugh]
Tanja: No... [continue laughing]. (FG 5: 3–13)

Associations with Greenlandic weather and changing seasons were present and prominent in all focus group discussions. In these immediate associations, climate change was equated with "weather" and objectified to become visible as a "warmer summer" and "the ice sheet, when it melts". Also, among the three middle-aged women of FG 2, who work in a school, a kindergarten and in banking, respectively, the close, personal observations of climate change were prevalent and absorbed the group's attention for much of the interview. Individual observations on changing seasons, drying areas of land, milder temperatures, less snow, stronger winds etc. were used to objectify climate change. Smith and Joffe's (2013: 26) distinction between *natural* and *unnatural* is omnipresent here, as interviewees juxtaposed ideas about "how nature is expected to behave" with observations of it behaving in unfamiliar and unpredictable ways. Participants in all five focus groups continued circling around the impacts of climate change in Nuuk and the rest of Greenland—how they experienced climate change in their local context. This process of objectifying through local images and examples was a main theme of all the interviews.

The continuous references to weather phenomena and their perceived changes aligns with previous findings from Wibeck (2014b: 209) and Smith and Joffe (2013: 212–22), who identified "melting polar ice caps, endangered polar bears, warmer weather, floods, and droughts", "melting ice", and "weather" as the dominant associations of their Norwegian and British interviewees. However, the present study's finding of strong contextual links to Greenland sets it apart from these earlier studies. While melting ice was mentioned, local experiences of weather change observed by the interviewees, themselves, or people they know took precedence in the interviews. Thus, personal experience appears to influence Nuuk's inhabitants' sense-making of climate change to a greater extent than what is concluded in studies of British and Norwegian laypeople's sense-making (Wibeck 2014b; Smith and Joffe 2013).

Another interesting aspect of the excerpt above is how humor was used to downplay the urgency and severity of climate change. Melting ice was laughed off as "alright". Along similar lines, Johannes elsewhere commented that in a couple of years, we will be able to come and visit "Costa del Nuuk" (FG 4: 183). I argue that this can be classified as a type of anchoring—inserting climate change in comical over-exaggerated representations, which makes it more tangible, less unsettling, and easier to talk about and make sense of. Bore and Reid (2014: 454) have previously explored how laughing at climate change can "promote active and positive engagement".

Certainty and Uncertainty with Regard to Climate Change's Anthropogenic Causes

Another significant finding was that all the interview participants reported having experienced climate change in Greenland, although they did not necessarily define this change as anthropogenic in origin. In interviews, the moderator consistently referred to the phenomenon as "climate change" but not as "anthropogenic" to avoid influencing participants' sense-making. This uncertainty regarding the question of attribution (see Chapter 7, Attribution Science) is illustrated by how climate change is discussed among the three women Jonna, Nivi, and Karen. When asked whether they believe in climate change, Jonna replied:

> Jonna: I do, really, I think there is change in so many things, also because you can also feel it on what you catch of whales and fish and other things. It is felt that, uhm, just now the cod has arrived much earlier than normal... (FG 2: 18)

These personal experiences, however, do not necessarily translate into a certainty that climate change has anthropogenic causes. To many interviewees, climate change is not necessarily 'human induced'. To some, it simply means a change in the climate. Thus, later in the interview, Jonna commented on how climate change is framed in the media as anthropogenic.

> Jonna: And then they used to say that it is the people themselves that are responsible for it. That we are—people. But I don't believe in that.

Several hundred years ago Europe was frozen. I certainly do not think that is the population's fault.
Karen: No.
... Moderator: What do the rest of you think of that?
Karen: I have also had that thought: that the atmosphere takes care of itself, the earth, right. (FG 2: 106–10)

Although Jonna has no doubt about the existence of a phenomenon she calls climate change, she does not define it as anthropogenic. Participants in other focus groups similarly expressed a wavering between anthropogenic climate change as *certain/uncertain*. Thus, Johannes declared his opinion:

Johannes: ...And the media, they describe it as manmade, right. But... (laughs), it bloody isn't all manmade. (FG 4: 4)

Diverging from Olausson's (2011: 287) finding that "human-induced" climate change was taken to be "common sense" among Swedish laypeople, this study does not identify a solidified social representation of climate change as manmade. This conclusion, however, ties into Bjørst's (2011: 239, author's translation) previous findings from Greenland, in which she identifies a "mistrust in science" and stronger belief in "locally formulated climate theories". It is, however, important to mention that the younger focus group participants clearly defined climate change as anthropogenic. Here, a common social representation of climate change as human induced seemed solidified to the extent that it was not necessary to discuss this further. In this study, it thus appears to be an age-dependent phenomenon: younger participants believe more strongly in anthropogenic climate change than do older participants.

Positive and Negative Impacts of Climate Change for the Self and the Other

Another major theme of the five focus group interviews was the evaluation of climate-change impacts as *positive/negative*. This was a divisive theme, dividing interviewees between worrying about the condition of nature and a pragmatic view focused on Greenland's economic development. A couple of interviewees drew this distinction in response to the interview's first question of what comes to mind when climate change is mentioned. Frederik, who works in an administrative position at a fish factory in Nuuk, began the interview as follows:

> Frederik: Uhm, I think possibilities (chuckles). Because it might be that if you live somewhere else than here, then it isn't possibilities, but what we see here is that things are changing a lot, I think, and that creates lots of possibilities for this country. (FG 1: 12)

When asked what he would write if asked to write an article about climate change later in the interview, Frederik answered:

> Frederik: The more the better, I guess. Then we can sit outside and my strawberries will feel better and we could grow some more potatoes. (FG 1: 110)

Other answers show that Frederik's evaluation of climate change as positive mainly relates to his profession and new possibilities for the fishery of e.g., cod, while an improved quality of life is typically added with a hint of humor. Other positive impacts of climate change mentioned in the focus groups are more climate tourists, agriculture, mining, longer summers, saving energy, and being able to sail wherever one wants. The positive evaluation of climate change as bringing "possibilities", as they were labelled independently by two interviewees, contradicts the dominant view in the media that climate change is purely negative. When asked which issues related to climate change he saw as most prominent in the media, Frederik interrupted:

> Frederik: Well, like I said, that is also why I said possibilities. Because everyone, almost everyone, focuses precisely on it being disasters this will bring, right? ... It is clear, if you live on some sort of island one and a half meters above the sea down in the Pacific Ocean, then reality might look a bit different. (FG 1: 67)

Thus, positive aspects of climate change were perceived as being in opposition to the dominant media representation of climate change as purely negative for local reasons. At the same time, Frederik's quote exemplifies how interviewees were reluctant to embrace a "victim" role. Here, Smith and Joffe's (2013: 23) distinction between *self* and *other* comes into play. Though some participants contemplated and worried about the negative impacts of climate change, the "self" was never regarded as a victim and an "other" very rarely as any form of "perpetrator" of climate change (Smith and Joffe 2013; Bjørst 2012). On the contrary, and as illustrated above, there was an inclination to cast *others*, such as the inhabitants of small islands threatened by sea level rise, as victims of climate change. This is also obvious in the following discussion between Frederik and his colleague Bjarne.

> Moderator: ...Maybe it also has something to do with media coverage, that there is somehow a counter reaction?
> Frederik: Yes, yes, exactly. That it becomes too much, that you think now it's all doomsday around that again.
> Bjarne: We can't just think about ourselves here in Greenland though.
> Frederik: No, that's also true.
> Bjarne: If it melts that much and it affects the oceans, right? You have to be very watchful about that. (FG 1: 111–16)

This focus group interview had only two participants, both working administratively at a fish factory in Nuuk. The contrasting view to media representations of climate change as inherently negative was highlighted again by Frederik's annoyance with "doomsday" reports. When the discussion had centered on the potential positive impacts of climate change in Greenland, Bjarne interrupted to remind his colleague that Greenland is not alone in the world. This highlights an additional distinction beyond those identified by Smith and Joffe: the distinction between the *local* and *global* impacts of climate change. This theme recurs in interviewees' discussions of their memories of media content outlined later in the analysis. In several interviews, the local level was accentuated through examples of the positive impacts of climate change in Greenland, showing how some interviewees did not adhere to the view that climate change is disastrous. The younger high school students also distinguished between *local* and *global*, albeit in a different form, as a demand for information about the de facto consequences of climate change for Greenland locally. This theme is also outlined below.

In other focus groups, climate change's negative impacts were more prevalent. In the following exchange, four high school students discussed whether climate change affects anyone they know.

> Ivalu: Hunters.
> Moderator: Hunters. How so?
> Ivalu: It's harder, I've heard.

Hansine: They can't go out sailing if it's that kind of weather, then they just have to stay at home.
Ivalu: And the weather is bad often.
David: Mmm. (FG 5: 66–71)

Hunters and fishermen were mentioned here and in other focus groups as a segment of the population that is especially vulnerable to climate change. Interviewees who mostly view climate change as positive recognized that it may pose more dire obstacles for inhabitants of other parts of the country. There was, however, a strong recognition that Greenlanders have adapted to changes in climate throughout history. Thus, respondents may view climate change as "worrying", but they do not use it to victimize themselves or other inhabitants of Greenland (FG 5: 112).

Environmental and Economic Sustainability: A Basic Distinction

The distinction between the *positive* and *negative* impacts of climate change underlines another distinction—between *environmental* and *economic sustainability*. Other researchers have touched on this theme, commenting on how Greenland is "warming to the idea of less snow and ice" due to its economic advantages (Nuttall 2009: 295). Ulrik outlined this view clearly in an evolving discussion with two other interviewees about how the country should behave in relation to new possibilities for mining.

Ulrik: Well, sustainable. The economy is simply also an important part of the sustainability, in my view. (FG 3: 140)

When asked if she sees any positive or negative impacts of climate change, Jonna, in another focus group, replied:

Jonna: Well, the positive is earnings. But who says that's the most important thing anyway? Yes, we do have a lot of these luxury things. Well, is that really what you need? I mean, should we be thinking luxury or should we be thinking about our nature? (FG 2: 147)

Although arguments on both ends of the *environment/economy* spectrum were contemplated in all focus groups, there was a noticeable tendency to highlight the positive impacts and possibilities of climate change in Greenland in the three focus groups comprised of people employed by an airline, a fish factory, and within governance and tourism. By contrast, the female interviewees studying or employed in the education and banking sectors interpreted climate change as predominantly negative. On the basis of this study, it is not possible to say whether gender, age, profession or even media use as variables correlate with sense-making of climate change as either positive or negative. Nonetheless, it is a finding worth contemplating that the interviewed men, working in industries that presumably are somewhat positively affected by climate change, viewed the phenomenon as predominantly positive.

Media and Personal Experience as Sources of Information

This section analyzes how the media and personal experiences might influence the interviewees' sense-making processes. It first looks at which information sources participants report using in the interviews and surveys. Second, it examines how they express and negotiate the trustworthiness of these sources.

The Media as a Source of Information on Climate Change

The media sources from which participants receive information about climate change varied within and across focus groups. In their survey responses, all participants except one reported hearing about climate change from national Greenlandic newspapers" "every week" or "every month". In all focus groups except FG 3, participants similarly reported hearing about climate change from "local Greenlandic newspapers" "every week" or "every month". The same is true for hearing about climate change from "TV". According to survey responses, traditional media such as national and local papers and TV comprise many of the participants' media use. Specific sources mentioned in interviews were the Greenlandic outlets *Sermitsiaq*, *AG* and KNR, the Danish newspaper *Politiken*, and television station TV 2, as well as the international news outlets CNN, *The New York Times*, *The Guardian*, *Siberian Times*, CBC, and the entertainment company Netflix.

In focus group interviews "anything from the local papers to the news on TV and, well, TV and written papers" to "Googling", "local" and "foreign media", and "radio" were mentioned as sources of information (FG 4: 48–50; FG 1: 52; FG 2: 104). Only the group of four high school students and the younger tourist guide Malthe stated that they mainly rely on articles from the internet, especially Facebook, coming from e.g., the Greenlandic news outlet Sermitsiaq. In their survey responses, however, Karen, Nivi, and Johannes also indicated that they hear about climate change through social media "every week", while Pavia does so "every day".

Memories of Media Content

Analyzing interviewees' memories of media coverage provides insight into media-related images through which they make sense of climate change. When the moderator scrutinized the examples brought up during the interviews, it once again became evident that interviewees frequently use images of Greenland to make sense of the phenomenon. Thus, even when speaking of mediated images, local Greenlandic objectifications of climate change were common. These are not necessarily connected specifically to Nuuk, but often illustrate the melting of Greenland's ice sheet. Thus, the interviewees often mentioned images from the news media including the "melting ice" and "melting polar ice caps" identified in Smith and Joffe's (2013) as well as Wibeck's (2014b) studies. The high school students David and Tanja recounted a time lapse video of the ice sheet's melting, while Ivalu recalled a video of two people watching an iceberg break off; "that was pretty crazy to see that it is so big. We live in Greenland but we don't see icebergs break off every day, you know (laughs)" (FG 5: 179).

In a second theme of memorized media images, interviewees recalled coverage of other populations affected by climate change through floods, which they recall as having made an impression on them. Both themes are illustrated well in an exchange between Nivi and Jonna.

> Nivi: I also think of a programme I watched, I think this year or last
> year. It was a recording from the ice sheet where it has become rivers.
> Water just streamed down. Just think how much water comes out every
> day. I have a hard time imagining it. You almost can't.
> Jonna: There is so much water coming out.
> Nivi: I wonder what's below the ice! And then I am thinking of this scene
> where I saw a flood, I don't quite remember where it was, but there were
> cars and everything is floating around, and men, people that had to be
> rescued. It is incomprehensible. (FG 2: 181–87)

Nivi and Jonna both remember media coverage showing the melting of the Greenlandic ice sheet. The images appeared to be vivid in their memory and seemed to affect them both. In the middle of this exchange, however, Nivi added an example of coverage of a flood in an unnamed place. A similar example is found in the exchange between Bjørn and Frederik. The latter remembered a news piece about an island state in the Pacific having bought new land to live on for when they become flooded. "That actually does make you think, okay, this really has a pretty big consequence for these people, right", says Frederik (FG 1: 140). His and Nivi's comments highlight how the participants, while mostly occupied with climate changes in their proximity, also objectify it through other people who experience its (negative) consequences. The remembered examples of media coverage related to Greenland usually objectify climate change through natural phenomena such as melting ice, while the examples related to other places that have made an impression mostly include direct negative consequences for people. Here, the distinction between *local* and *global* is drawn again, as interviewees supply examples of how other places are more affected by climate change. The distinction was present yet again as the focus group of high school students discussed how they do not know what will happen to Greenland when the ice melts.

> Ivalu: We just know that Denmark will sink.
> Hansine: We just know that Denmark will become flooded (laughs)
> Ivalu: Then they will all have to move here.
> David: Yeah.
> Ivalu: Then we'll get the queen (all laugh). (FG 5: 204–10)

Here, a distinction between Greenland and Denmark is drawn, describing how, even though climate change is vivid to interviewees, its most severe consequences are cast as affecting someone else.

Negotiating the Trustworthiness of Media and Personal Experiences

Exploring the information sources listed by participants in their survey responses and through focus group interviews, the main finding is that media and personal experience both appear to be important information sources. This is exemplified in the following exchange among four high school students, when asked where they get information on climate change:

> Ivalu: The media.
> Moderator: The media?
> Ivalu: Yes.
> David: Yeah, the media.
> Tanja: Yes.
> Moderator: Is it mostly media it comes from, do you think?
> Ivalu: Yes, but also—for me also from people on the street, like local fishermen and hunters that talk about it, and old people.
> Hansine: Mmm.
> Ivalu: And for example my grandmother, everything like that.
> Moderator: Yeah. What do they say?
> Ivalu: How cold it was in the old times, before we were born. Back when the snow really squeaked in the winter and it was really cold so your ears turned red, and all that kind of stuff that my mum used to talk about.
> David: Mmm.
> Hansine: Yes. (FG 5: 80–92)

The above exchange shows how media and first-hand sources interact and are negotiated. Several exchanges across the different focus groups reveal how the interviewees alternate between using the media and personal experiences to make sense of climate changes. This was also exemplified in how some interviewees counter negative media representations using personal and locally contextualized evaluations of climate change as positive.

In the previous section, the four high school students started by mentioning the media as their primary source of information, and, after an intervention by the moderator, added elders, hunters and fishermen as sources. The following exchange indicates that this switch also implies that they trust first-hand sources more. When asked whether they trust the media or first-hand sources more, they replied:

> Ivalu: I trust old people more (laughs). Because they know what they are talking about.
> David: Yeah, that's right.
> Hansine: Yes.
> Tanja: Mmm.
> Ivalu: They really know what they're talking about.
> Moderator: Yeah. So in comparison with an article, then you would trust most..?
> Hansine: The elders. (FG 5: 151–59)

The question about which sources are more trustworthy was negotiated through the interview, and thus is not a set or fossilized representation. When the moderator suggested that elders might just have a different perspective than the media, Ivalu replied, "I trust my grandmother more than I trust the media" (FG 5: 161). In general, although the media was mentioned as a source of information in all focus groups, its trustworthiness was scrutinized more than that of personal experiences and sources.

In their survey responses, all female participants except Jonna ranked "conversations with friends, family, colleagues etc." as the source they rely on most for climate change information. But all three women in FG 2 also included either "local" or "national Greenlandic newspapers" in their list of trustworthy sources. These survey responses align closely with the three women's collective sense-making of climate change through objectifications of local changes in weather in their focus group interview. Interestingly, in another part of the survey respondents (including the three airline employees and the Danish government officials Gorm and Ulrik) ranked "international newspapers" among the most reliable sources. This matches the focus in their discussions, which were more preoccupied with the positive and economically beneficial sides of climate change than its negative environmental consequences.

Media Representations Related to Personal Experience

This section assesses how interviewees negotiated and made sense of contradictions between mediated and personal experiences of climate change.

Criticism of "Doomsday" Coverage

First, the study explored what the interviewees believe the media focus on in their climate change coverage. A main theme in interviewees' responses to this question was that the media focused too much on the negative aspects of climate change or incited fear and alarmism through their coverage. A clear example is Frederik's persistent highlighting of the positive possibilities that climate change poses in Greenland, which he finds missing from the media's "doomsday" coverage, as outlined above (FG 1: 112, 67). The theme is also clear in Malthe's response to how he views media reporting on climate change.

> Malthe: Fear mongering, I think. Honestly, I think so. It's very, like, now you should only shower once a week and cycle to work or take the bus at least. Maybe the rest of you have a different view?

Malthe expressed the view that the media are overly focused on the catastrophic elements of climate change. The same view was expressed in several focus group interviews. For instance, Karen reported that the media sometimes "blows it out of proportion a little too much" (FG 2: 145). This type of "emotional fatigue" has been detected in other studies by, e.g., Olausson (2011: 292), whose focus group participants were "highly critical of the commercial conditions of the news media that generate this type of journalism". The finding that emotional appeals, especially to fear, may have the opposite effect to increasing concern about climate change is supported in several other studies (Ryghaug, Sorensen, and Naess 2011; Wolf and Moser 2011; O'Neill and Nicholson Cole 2009).

A criticism of the media's lack of alarmism set the high school students apart from the other interviewees. Their emphasis on how the media does not take climate change seriously enough is exemplified in the following exchange, when asked what they generally think about how the media covers the issue.

> Ivalu: It's very mild.
> David: Yes, it's mega mild. Really mild.
> Tanja: Kindergarten.
> Ivalu: It is not so much, we need to do something now and it is serious and it affects a lot of people, it affects the whole earth.
> David: Mhm.
> Moderator: So by "very mild", you mean—what do you mean by "mild"?
> Tanja: Kindergarten mild.
> David: I mean it won't affect anyone at all. That mild. (FG 5: 264–71)

Similar expressions of dissatisfaction with how seriously climate change is taken in the media were not found in any other focus group. As seen in the passage above, the students even collectively showed disdain for how "mildly" the media treats the topic of climate change, calling it "kindergarten mild".

Lack of Local Voices

The high school students also criticized the *lack of a local focus* on climate change in the media.

> David: The media, they cover Denmark's consequences but there is not really anything about covering our consequences, if something should happen. (FG 5: 183)

This social representation of how the consequences of climate change for Greenland are missing from the media coverage seemed to develop over the course of the conversation, as Ivalu's reply indicates, "yeah, that's true actually... Yeah, wow" (FG 5: 184–86). Once the idea was conceived, however, it reoccurred several times throughout the interview. Thus, when asked what they would write about if they were to write an article about climate change, the students reply:

> Ivalu: Listen to us. And to the elders. I mean listen to the locals. Listen to us who live here.
> David: Yes.
> Ivalu: Not necessarily Vittus but.. (all laugh)
> Moderator: ...Who would you interview then?

Ivalu: Someone who is reasonable.
David: It's the hunters, I think. The hunters and the locals.
Ivalu: Also the ones in smaller towns, settlements.
Tanja: Yeah. (FG 5: 329–40)

This idea seems linked to how the students claim to trust elders, fishermen, and hunters from Greenland more than the media on questions related to climate change. It is an example of precisely how focus group participants relate personal experiences to media representations and at times identify dissonances between the two. Here, the students employed the accounts they have collected through personal interactions to critique how media representations exclude important voices. This theme was only peripherally touched on in other focus groups, for example when Pavia declared that he thinks climate conferences are a waste of time, and should instead take into account the experiences of "real nature people." (FG 4: 75)

This finding touches on the extent to which the media are seen to represent the local, contextual actualizations of climate change around the world. It aligns with the critique from other focus groups that find positive Greenlandic perspectives missing in, particularly non-Greenlandic, media accounts of the issue. It is interesting to note how the students highlighted personal, first-hand accounts and requested a stronger focus on local, Greenlandic accounts of climate change in the media, while at the same time highlighting the severity of climate change. In other focus groups, local perspectives were mainly used to exemplify the positive effects of climate change in Greenland and to criticize media representations for focusing only on its negative effects. This could be a generational difference, as the high school students are the group that most separated themselves from the four other focus groups in terms of their greater certainty regarding anthropogenic climate change and sense-making of the phenomenon as predominantly negative. This could, however, also be related to many other variables such as media use. Unfortunately, no survey responses were obtained from the high school students to investigate these issues.

Another negotiation of how local Greenlandic experiences of climate change are mirrored in media representations was present in the tourist guide Malthe's proclamation that Greenland is sometimes represented as a victim "in others' argumentation". When asked to elaborate on this point, he replied:

> Malthe: I think you often experience Greenland as this big ice cube that's just melting and melting, melting and the water levels are rising and rising and rising. That's mostly what I meant with that—in a way becoming a victim. Or positioned in this kind of victim role. Whereas if you ask the Greenlanders themselves, like me for example, then it's a completely different picture one has. (FG 3: 182)

Malthe's comments outline a dissonance between a common media representation of climate change and the local, contextual experiences of himself and other Greenlanders. In a sense, he distances himself from the common objectifying images of climate change, namely "melting ice" and rising "sea levels"—which studies have found are the most common images of climate change for many, at least Western, people (Wibeck 2014b; Smith and Joffe 2013; Leiserowitz 2006). Malthe's comments illuminate another way through which interviewees related media representations to their personal experiences, thus negotiating established social representations and creating new ones in the process. It very explicitly underlines the need to include varied, local, contextualized experiences of even a phenomenon as global as climate change.

Climate Conferences and Politicians' Visits

Another media focus, according to the interviewees, is stories about climate conferences and "high-profile climate tourists" visiting Greenland, as the participant Gorm calls them. This was also exemplified in Ulrik's observation on media stories below.

> Ulrik: Well, specifically, I would say, that when some big shot or other from somewhere comes up here to see it. And then it's only to hear or read what they all say. (FG 3: 62)

These two types of stories, high-profile politicians at climate conferences or passing through Greenland, were brought up in several interviews. Interviewees also mentioned climate conferences as the most cited sources in climate change news coverage when asked directly. They discussed how climate conferences are held and widely covered, but expressed no particular trust in their effect. On the contrary, Malthe evoked a pejorative image of "grown up people with a long education honestly sitting and just talking about the weather" (FG 3: 26). The political negotiations on climate change were thus another representation by which interviewees made sense of the issue.

Discussion

This study explores how fifteen inhabitants of the world's northernmost capital, Nuuk, make sense of climate change through media exposure and personal experience. The three distinctions outlined in Smith and Joffe's (2013) study (*natural/unnatural, certainty/uncertainty and self/other*) were identified in the focus group discussions, and were complemented by three additional distinctions based on the empirical data (*local/global, positive/negative,* and *environment/economy*). Together, they represent the main points of contention or organizing "opposites" that the fifteen interviewees used to make sense of climate change (Olausson 2011).

Natural/Unnatural

As is clear in their myriad objectifications of climate change, the distinction between *natural* and *unnatural* permeated interviewees' sense-making of climate change. Smith and Joffe (2013: 26) describe this distinction as the juxtaposition of "how nature is expected to behave" with observations of it behaving in unfamiliar or unpredictable ways. Climate change was continuously represented as "weather", which was perceived as stormier, more unreliable, warmer, etc. in comparison to before. This naming mechanism was also identified among participants in Olausson's (2011: 289) study, allowing her to conclude that "the abstract and intangible risk acquire[d] familiar and comprehensible characteristics".

Certainty/Uncertainty

Notably, none of the interviewees doubted the existence of climate change. The findings of the study, however, illuminate how perceptions of changes in the Greenlandic climate did not necessarily translate into a certainty with regard to anthropogenic climate change. Interviewees repeatedly discussed whether climate change is manmade by references to how "you can't do anything against nature" and the notion that climate change should not be blamed on anyone. A social representation of climate change as manmade was not solidified among most participants. The distinction between *certainty* and *uncertainty* was thus an important sense-making device. In contrast, Olausson's (2011: 287) study of Norwegian laypeople's sense-making showed a high degree of certainty that climate change is manmade.

From Self/Other to Local/Global

The study's findings showed how the study participants did not use the *self/other* distinction to portray an "other" as the perpetrator of climate change. Contrary to Smith and Joffe's (2013) findings, the participants in Nuuk advocated against blaming anyone for climate change. Instead, they used the *self/other* distinction to explain how the impacts of climate change were felt more strongly by others—such as islanders in the Pacific Ocean and hunters in the more northern regions of Greenland. This finding confirms Ryghaug, Sorensen, and Naess' (2011: 785) description of interviewees' interpretation of climate change as "distant in time and space", thereby giving it a "less stressful" meaning. This distinction parallels the more specific difference between *local* and *global* identified in the present study. To some interviewees, the realities of climate change on a global scale should of course not be ignored, but on a local level, the impacts were seen as mostly positive. Interviewees also observed differences between the focus in local vs. global media coverage. The distinction between *local* and *global* adds an important extra layer to the distinction of *self* versus *other*. In essence, it highlights how interviewees are able to make sense of climate change, as both a global phenomenon with mainly negative impacts, and on the local level, as the "possibility" of pleasantly warmer temperatures and new economic potential.

Positive/Negative

Throughout the focus groups, a new *positive/negative* distinction was identified. While there were "romanticized recollections" of snow up to one's head in the past, romanticized imaginations of a future, warmer "Costa del Nuuk" were just as prevalent in the interviews (Smith and Joffe 2013). Positive impacts of climate change in Nuuk were often advanced in opposition to fear, inciting media representations of climate change as purely negative. Other studies, such as Ryghaug, Sorensen, and Naess' (2011: 785), have illuminated how Norwegian citizens reason against the severity in media representations in order to "diminish the risks" of global warming and give it a "less agonizing meaning". In this study, I would argue that the emphasis on the positive impacts of climate change also serves to underline one's own, localized experiences of climate change when these are not recognized in media representations. Participants who experienced positive effects of climate change, such as through their profession, criticized the media for not sufficiently portraying these positive effects. Thus, the distinction between evaluating climate-change impacts as positive or negative was a central theme of the interviews, and a unique finding of this study. It illuminates a dissonance between personal experiences and media representations of climate change within a particularly climate-exposed population. It is interesting to note how this finding differs from the results of the nationally representative survey conducted by Minor et al. (2019), in which respondents saw climate change and its impacts in Greenland as mainly negative. A possible explanation could be their different methodologies, but also the time period that separates them is interesting to observe as climate change has received rising global political attention. The deviating findings clearly underline the need for further research in this area.

Environment/Economy

A classic distinction in ecological debates is the tension between the *environment* and the *economy*. Previous studies connect this distinction to that between *positive/negative*. Bjørst's (2011: 243, author's translation) finding that her informants in Greenland stressed "possibilities as well as problems" is thus confirmed with the findings among the laypeople of Nuuk. It is interesting to note that the participants closest to the economy end of the *environment/economy* continuum were first government officials and second employed in the tourism and fisheries industries, which can be seen to be positively affected by climate change. Those aligned with the environment end of the continuum were high school students and women working in "unaffected" positions in the education and banking sectors. This finding can be connected to Olausson's (2011: 294) assumption that when social representations do not match our own "experiences, values, and opinions", they are "likely to be transformed to resemble already established and familiar social representations, or even to be rejected to avoid any cognitive or emotional dissonance". Future, larger-scale quantitative studies could focus, for example, on how inhabitants' ages, professions, and educational backgrounds influence the sense-making of climate change in Greenland. Participants did not state that their preoccupation with the positive impacts of climate change emanated from or was inspired by media coverage—quite the contrary, these points were brought up in sharp contrast to mainstream media coverage. Nuttall, however, has argued that Greenlanders increasingly look more positively at climate change as politicians encourage the public to "think positively about the opportunities that climate change is bringing" (2009:47; 2008: 295).

Personal Experience and the Media as Information Sources

A crucial finding of this study is the participants' reliance on personal experience in their sense-making of climate change. In objectifying climate change, interviewees drew to a great extent on unmediated, personal images. For example, when asked whether she believes in climate change, a young high school student replied, "I mean, I have experienced it my whole life" (FG 5: 35). The importance that participants placed on anecdotes and stories heard from elders and fishermen reveal how interpersonal communication permeates participants' sense-making. In line with this, several previous studies have shown how "respondents make use of everyday experiences of the weather" to make sense of climate change in Sweden, Norway, and Britain (Olausson 2011; Ryghaug, Sorensen, and Naess 2011; Lorenzoni and Hulme 2009; Whitmarsh 2008). Interviewees also drew on mediated objectifications in their sense-making of climate change. These were often strongly connected to Greenland. These objectifications included vivid images of the melting ice sheet and rivers formed by thawing and large break offs of icebergs, as seen in videos on the internet. Climate change was also objectified through images from other parts of the world, such as island states in the Pacific. The mediated objectifications of climate change among Nuuk's inhabitants therefore align somewhat with those identified among British and Swedish respondents in other studies (Wibeck 2014b; Smith and Joffe 2013). In those studies, "melting ice", "weather", "melting polar ice caps, endangered polar bears, warmer weather, floods, and droughts" dominated interviewees' objectifications of climate change (Wibeck 2014b: 209; Smith and Joffe 2013: 21). This is an interesting finding, which suggests that a global set of images related to climate change has diffused through the media to Norwegians as well as Brits, Swedes, and Greenlanders.

Opposing Media Representations

As the analysis and discussion have so far revealed, media exposure and personal experiences of climate change form parts of a complex reception and sense-making process. Stamm, Clark, and Eblacas (2000: 220) conclude that studies in this field show that media coverage is "at least partly responsible for focusing people's attention on environmental problems". The current study demonstrates that residents of Nuuk also rely heavily on the media for information on climate change. This does not, however, mean that Nuuk's inhabitants uncritically absorb media representations. Just as Olausson (2011: 294) concludes, media information was "negotiated and remolded in conversations and discussions with other people" throughout the course of the focus group interviews.

This study's analysis has illuminated how participants were able to critically reflect on the information they received from the media. The most frequent criticisms of the media were related to the prevalence of emotional appeals and incitement of fear. By contrast, a minority of younger interviewees criticized the media for being too mild in its climate change reporting, and failing to communicate the urgency of the matter. Other participants missed local, Greenlandic accounts of climate change and its consequences. Overall, the participants from Nuuk cannot be seen as "passive recipients of powerful messages", as they actively negotiated media representations (Jensen and Rosengren 1990).

Olausson (2011: 295) describes the role of the media in shaping sense-making of climate change as that of an "agenda-setter", in the sense that it establishes an "overall framework" that is then "gradually filled with various elements, including personal and collectively deliberated experiences". This account aligns well with the findings from Nuuk, showing how interviewees draw on personal observations to make sense of mediated concepts of climate change such as the threat of climate change. According to Olausson 's (2011: 295) use of the concept, agenda-setting cannot be reduced to focusing people's attention or "making people talk about climate change", but also encompasses "setting the limits for viable ways of talking about this global risk in terms of causes, consequences, and responsibility for solutions". In this sense, although they may not be directly visible, established social representations of climate change, perpetuated by the media, set limits on sense-making of the phenomenon. Although this study has convincingly shown that the social representations of Nuuk's inhabitants oppose media representations of climate change as purely negative, their reasoning about the positive aspects of climate change may also be grounded in media representations. Indeed, a couple of interviewees noted that the positive aspects of climate change are more prevalent in national Greenlandic media.

Conclusion

This chapter has explored how inhabitants of Greenland's capital city make sense of climate change and its impacts through media and personal experiences. Social representations theory was used to analyze the processes through which the scientific phenomenon was made tangible and comprehensible by fifteen inhabitants of various ages, genders, professions and backgrounds.

In order to ensure a meaningful adaptation to the climatic changes experienced in Nuuk, we should learn from its inhabitants' sense-making. This entails including their locally particular, geographically and culturally specific representations of climate change—for example, bringing Greenlanders' positive experiences of climate change, although contradictory to dominant narratives—in broader media representations of climate change. These local representations challenge both international research and media to "create space for multiple and under-represented voices on the experience of climate change" (Farbotko and Lazrus 2012: 383). Otherwise, the voices and interests of populations at the heart of climate (crisis) discourses are in danger of being misrepresented or excluded. Thus, paying attention to local discourses on climate change is a crucial part of understanding the issue, its impacts, and ways to mitigate and adapt to it.

Future studies should incorporate larger national and cross-national perspectives of public views on climate change in Greenland and other parts of the Arctic region. The survey incorporated in this study can only illuminate the particular media use of the interviewees. A larger-scale study of climate change-related media use could further explore some of the perspectives analyzed in this study regarding how media use influences sense-making of climate change in Greenland. A study of national Greenlandic media coverage of climate change should also be undertaken to investigate whether the positive evaluations found in this study are grounded in the images and evaluations of national media coverage. Finally, future case studies should continue to explore how climate change is made sense of in local contexts around the world. While the Greenland that is "melting away" comprises an internationally recognized representation, its anticipated future capital "Costa del Nuuk" is—so far—less well known.

References

Ackrén, Maria, and Uffe Jakobsen. 2015. "Greenland as a Self-Governing Sub-National Territory in International Relations: Past, Current and Future Perspectives", *Polar Record*, 51.4: 404–12, https://doi.org/10.1017/s003224741400028x

Arctic Monitoring and Assessment Programme. 2012. *Arctic Climate Issues 2011: Changes in Arctic Snow, Water, Ice and Permafrost* (Oslo, Norway, Ottawa, Ontario: AMAP), https://www.amap.no/documents/doc/arctic-climate-issues-2011-changes-in-arctic-snow-water-ice-and-permafrost/129

Arlt, Dorothee, Imke Hoppe, and Jens Wolling. 2011. "Climate Change and Media Usage: Effects on Problem Awareness and Behavioural Intentions', *International Communication Gazette*, 73.1-2: 45–63, https://doi.org/10.1177/1748048510386741

Bjørst, Lill R. 2008. "Grønland og den dobbelte klimastrategi", *Økonomi Og Politik*, 81.4: 26–37.

——. 2011. "Arktiske diskurser of klimaforandringer i Grønland: Fire (post) humanistiske studier" (PhD thesis, Københavns Universitet).

——. 2012. "Climate Testimonies and Climate-crisis Narratives. Inuit Delegated to Speak on Behalf of the Climate", *Acta Borealia*, 29.1: 98–113, https://doi.org/10.1080/08003831.2012.678724

Bore, Inger-Lise K., and Grace Reid. 2014. "Laughing in the Face of Climate Change?: Satire as a Device for Engaging Audiences in Public Debate", *Science Communication*, 36.4: 454–78, https://doi.org/10.1177/1075547014534076

Bravo, Michael T. 2009. "Voices from the Sea Ice: The Reception of Climate Impact Narratives", *Journal of Historical Geography*, 35.2: 256–78, https://doi.org/10.1016/j.jhg.2008.09.007

Brüggemann, Michael, Fenja De Silva-Schmidt, Imke Hoppe, Dorothee Arlt, and Josephine B. Schmitt. 2017. "The Appeasement Effect of a United Nations Climate Summit on the German Public", *Nature Climate Change*, 7.11: 783–87, https://doi.org/10.1038/nclimate3409

Cabecinhas, R., A. Lázaro, and A. Carvalho. 2008. "Media Uses and Social Representations of Climate Change", in *Communicating climate change: Discourses, mediations and perceptions*, ed. by Anabela Carvalho (Braga: Centro de Estudos de Comunicação e Sociedade (CECS)), pp. 170–89.

Davenport, Coral, Josh Haner, Larry Buchanan, and Derek Watkins. 2015. "Greenland is Melting Away", *The New York Times*, 27 October, https://www.nytimes.com/interactive/2015/10/27/world/greenland-is-melting-away.html

David, Matthew, and Carole D. Sutton. 2011. *Social Research: An Introduction* (Los Angeles: Sage).

Encyclopaedia Britannica. 2017. "Nuuk", *Encyclopaedia Britannica*, https://www.britannica.com/place/Nuuk

Farbotko, Carol, and Heather Lazrus. 2012. "The First Climate Refugees? Contesting Global Narratives of Climate Change in Tuvalu", *Global Environmental Change*, 22.2: 382–90, https://doi.org/10.1016/j.gloenvcha.2011.11.014

Gill, Nicholas. 2015. "Where is the World's Most Remote City?", *The Guardian*, 19 August, https://www.theguardian.com/cities/2015/aug/19/where-worlds-most-remote-city

Government of Greenland. *Economy and Industry in Greenland*, https://naalakkersuisut.gl/en/About-government-of-greenland/About-Greenland/Economy-and-Industry-in-Greenland

——. *Grønlandsk territorialt forbehold for klima-aftalen*, https://naalakkersuisut.gl/da/Naalakkersuisut/Nyheder/2016/04/190416-Klimaaftale

——. *Politics in Greenland*, http://naalakkersuisut.gl/en/About-government-of-greenland/About-Greenland/Politics-in-Greenland

Greene, Jennifer C., Valerie J. Caracelli, and Wendy F. Graham. 1989. "Toward a Conceptual Framework for Mixed-Method Evaluation Designs", *Educational Evaluation and Policy Analysis*, 11.3: 255–74, https://doi.org/10.3102/01623737011003255

Hall, Dorothee K., Richard S. Williams, Kimberly A. Casey, Nicolo E. Di Girolamo, and Zhengming Wan. 2006. "Satellite-Derived, Melt-Season Surface Temperature of the Greenland Ice Sheet (2000–2005) and its Relationship to Mass Balance", *Geophysical Research Letters*, 33.11: 787, https://agupubs.onlinelibrary.wiley.com/doi/full/10.1029/2006GL026444

Harvey, Chelsea. 2016. "Climate Change: Greenland Loses a Trillion Tonnes of Ice in Four Years as Melting Rate Triples", 21 July, https://www.independent.co.uk/environment/climate-change-global-warming-greenland-ice-melting-rate-sea-levels-rise-a7147846.html

Höijer, Birgitta. 2010. "Emotional Anchoring and Objectification in the Media Reporting on Climate Change", *Public Understanding of Science*, 19.6: 717–31, https://doi.org/10.1177/0963662509348863

Holm, L. K. 2010. "Sila-Inuk: Study of the Impacts of Climate Change in Greenland", in *SIKU: Knowing Our Ice Documenting Inuit Sea Ice Knowledge and Use*, ed. by Igor Krupnik, et al. (Dordrecht, Heidelberg: Springer), pp. 145–60, https://doi.org/10.1007/978-90-481-8587-0_6

IPCC. 2014. *Climate Change 2014 Synthesis Report: Contribution of Working Groups I, II and III to the Fifth Assessment Report of the Intergovernmental Panel on Climate Change* (Geneva: IPCC), https://www.ipcc.ch/report/ar5/syr/

Jensen, Klaus B., and Rasmus Helles. 2015. *Spørgsmål til spørgeskemaundersøgelser om danskernes medievaner 2014*, https://mediernesudvikling.slks.dk/fileadmin/user_upload/dokumenter/medier/Mediernes_udvikling/2015/Specialrapporter/KU_Mediebrug/Danskernes_mediebrug_2014_Rapport.pdf

Jensen, Klaus B., and Karl E. Rosengren. 1990. "Five Traditions in Search of the Audience", *European Journal of Communication*, 5.2: 207–38, https://doi.org/10.1177/0267323190005002005

King, Nigel, and Christine Horrocks. 2010. *Interviewing in Qualitative Research* (London: Sage).

Leduc, Timothy B. 2007. "Sila Dialogues on Climate Change: Inuit Wisdom for a Cross-Cultural Interdisciplinarity", *Climatic Change*, 85.3-4: 237–50, https://doi.org/10.1007/s10584-006-9187-2

Leiserowitz, Anthony. 2006. "Climate Change Risk Perception and Policy Preferences: The Role of Affect, Imagery, and Values", *Climatic Change*, 77.1-2: 45–72, https://doi.org/10.1007/s10584-006-9059-9

Lorenzoni, Irene, and Mike Hulme. 2009. "Believing is Seeing: Laypeople's Views of Future Socio-Economic and Climate Change in England and in Italy", *Public Understanding of Science*, 18.4: 383–400, https://doi.org/10.1177/0963662508089540

Lorenzoni, Irene, and Nick F. Pidgeon. 2006. "Public Views on Climate Change: European and USA Perspectives", *Climatic Change*, 77.1-2: 73–95, https://doi.org/10.1007/s10584-006-9072-z

Lüthcke, Scott B., H. Jay Zwally, Waleed Abdalati, David D. Rowlands, and Richard D. Ray. 2006. "Recent Greenland Ice Mass Loss by Drainage System from Satellite Gravity Observations", *Science*, 314.5803: 1286–89, https://doi.org/10.1126/science.1130776

Marková, Ivana. 2003. *Dialogicality and Social Representations: The Dynamics of Mind* (New York: Cambridge University Press).

Martello, Marybeth L. 2004. "Global Change Science and the Arctic Citizen", *Science and Public Policy*, 31.2: 107–15, https://doi.org/10.3152/147154304781780082

——. 2008. "Arctic Indigenous Peoples as Representations and Representatives of Climate Change", *Social Studies of Science*, 38.3: 351–76, https://doi.org/10.1177/0306312707083665

Metag, Julia, Tobias Füchslin, and Mike S. Schäfer. 2015. "Global Warming's Five Germanys: A Typology of Germans' Views on Climate Change and Patterns of Media Use and Information", *Public Understanding of Science*, 26.4: 434–51, https://doi.org/10.1177/0963662515592558

Minor, Kelton, Gustave Agneman, Navarana Davidsen, Nadine Kleemann, Ulunnguaq Markussenn, et al. 2019. *Greenlandic Perspectives on Climate Change (2018–2019): Results from a National Survey* (University of Greenland and University of Copenhagen: Kraks Fond Institute for Urban Research), https://kraksfondbyforskning.dk/wp-content/uploads/2019/08/2019_08_11_Greenlandic_Perspectives_Climate_Change_Final_Report_Reduced.pdf

Moloney, Gail, Zoe Leviston, Timothy Lynam, Jennifer Price, Samantha Stone-Jovicich, et al. 2014. "Using Social Representations Theory to Make Sense of Climate Change: What Scientists and Nonscientists in Australia Think", *Ecology and Society*, 19.3, https://doi.org/10.5751/es-06592-190319

Moscovici, Serge. 1984a. "The Myth of the Lonely Paradigm: A Rejoinder", *Social Research*, 51.4: 939–67, https://www.researchgate.net/publication/286387562_The_myth_of_the_lonely_paradigm_A_rejoinder

——. 1984b. "The Phenomenon of Social Representations", in *Social Representations*, ed. by R. M. Farr and Serge Mosovici (Cambridge, UK: Cambridge University Press), pp. 3–69.

——. 2000. *Social Representations: Explorations in Social Psychology* (Cambridge, UK: Polity)

Nuttall, Mark. 2008. "Climate Change and the Warming Politics of Autonomy in Greenland", *Indigenous Affairs*, 1–2: 44–51.

——. 2009. "Living in a World of Movement: Human Resilience to Environmental Instability in Greenland", in *Anthropology and Climate Change: From Encounters to Actions*, ed. by Susan A. Crate and Mark Nuttall (Walnut Creek, CA: Left Coast Press), pp. 292–310.

Olausson, Ulrika. 2011. "'We're the Ones to Blame': Citizens' Representations of Climate Change and the Role of the Media", *Environmental Communication*, 5.3: 281–99, https://doi.org/10.1080/17524032.2011.585026

O'Neill, Saffron, and Sophie Nicholson-Cole. 2009. "'Fear Won't Do It': Promoting Positive Engagement With Climate Change Through Visual and Iconic Representations", *Science Communication*, 30.3: 355–79, https://doi.org/10.1177/1075547008329201

Rasmussen, Rasmus O. 2016. *Grønland*, https://denstoredanske.lex.dk/Grønland

Ravn-Højgaard, Signe, Naja Paulsen, Mariia Simonsen, Naimah Hussain and Ida Willig. 2018. *Tusagassiuutit 2018—en kortlægning af de grønlandske medier*, https://uni.gl/media/4352533/dktusagassiuutit2018rapport.pdf

Rischel, J. 2016. *Grønland—sprog*, 23 June, https://denstoredanske.lex.dk/Gr%C3%B8nland_-_sprog?utm_source=denstoredanske.dk&utm_medium=redirectFromGoogle&utm_campaign=DSDredirect

Roosvall, Anna, and Matthew Tegelberg. 2012. "Misframing the Messenger: Scales of Justice, Traditional Ecological Knowledge and Media Coverage of Indigenous People and Climate Change", in *Media Meets Climate: The Global Challenge for Journalism*, ed. by Elisabeth Eide and Risto Kunelius (Göteborg: NORDICOM), pp. 297–312

Ryghaug, Marianne, Knut Holtan Sorensen, and Robert Naess. 2011. "Making Sense of Global Warming: Norwegians Appropriating Knowledge of Anthropogenic Climate Change", *Public Understanding of Science*, 20.6: 778–95, https://doi.org/10.1177/0963662510362657

Schäfer, Mike S. 2015. "Climate Change and the Media", in *International Encyclopedia of the Social and Behavioral Sciences*, ed. by James D. Wright, 2nd edn (Amsterdam: Elsevier), pp. 853–59, https://doi.org/10.1016/b978-0-08-097086-8.91079-1

Schäfer, Mike S., and Inga Schlichting. 2014. "Media Representations of Climate Change: A Meta-Analysis of the Research Field", *Environmental Communication*, 8.2: 142–60, https://doi.org/10.1080/17524032.2014.914050

Schultz-Lorentzen, H., and Rasmus O. Rasmussen. 2012. *Nuuk Kommune*, https://denstoredanske.lex.dk/Nuuk?utm_source=denstoredanske.dk&utm_medium=redirectFromGoogle&utm_campaign=DSDredirect

Smith, Nicholas, and Helene Joffe. 2013. "How the Public Engages with Global Warming: A Social Representations Approach", *Public Understanding of Science*, 22.1: 16–32, https://doi.org/10.1177/0963662512440913

——. 2009. "Climate Change in the British Press: The Role of the Visual", *Journal of Risk Research*, 12.5: 647–63, https://doi.org/10.1080/13669870802586512

Stamm, Keith R., Fiona Clark, and Paula R. Eblacas. 2000. "Mass Communication and Public Understanding of Environmental Problems: The Case of Global Warming", *Public Understanding of Science*, 9.3: 219–37, https://doi.org/10.1088/0963-6625/9/3/302

Statistikbanken. "Befolkningen i Nuuks bydele 1. juli 2011–2018", http://bank.stat.gl/pxweb/da/Greenland/Greenland__BE__BE01__BE0140/BEXSTMNUK.PX/?rxid=BEXSTMNUK25-07-2019%2017:15:42

Taddicken, Monika. 2013. "Climate Change from the User's Perspective", *Journal of Media Psychology: Theories, Methods, and Applications*, 25.1: 39–52, https://doi.org/10.1027/1864-1105/a000080

Tejsner, Pelle. 2013. "Living with Uncertainties: Qeqertarsuarmiut Perceptions of Changing Sea Ice", *Polar Geography*, 36.1-2: 47–64, https://doi.org/10.1080/1088937x.2013.769282

Whitmarsh, Lorraine. 2008. "What's in a Name?: Commonalities and Differences in Public Understanding of 'Climate Change' and 'Global Warming'", *Public Understanding of Science*, 18.4: 401–20, https://doi.org/10.1177/0963662506073088

Whitmarsh, Lorraine, Irene Lorenzoni, and Saffron O'Neill. 2011. *Engaging the Public with Climate Change: Behaviour Change and Communication* (Hoboken: Earthscan)

Wibeck, Victoria. 2014a. "Enhancing Learning, Communication and Public Engagement about Climate Change—Some Lessons from Recent Literature", *Environmental Education Research*, 20.3: 387–411, https://doi.org/10.1080/13504622.2013.812720

——. 2014b. "Social Representations of Climate Change in Swedish Lay Focus Groups: Local or Distant, Gradual or Catastrophic?", *Public Understanding of Science*, 23.2: 204–19, https://doi.org/10.1177/0963662512462787

Wolf, Johanna, and Susanne C. Moser. 2011. "Individual Understandings, Perceptions, and Engagement with Climate Change: Insights from In-Depth

Studies across the World", *WIREs Climate Change*, 2.4: 547–69, https://doi.org/10.1002/wcc.120

Yin, Robert K. 2014. *Case Study Research: Design and Methods*, 5th edn (Los Angeles: Sage)

Zwally, H. Jay, Waleed Abdalati, Tom Herring, Kristine Larson, Jack Saba, et al. 2002. "Surface Melt-Induced Acceleration of Greenland Ice-Sheet Flow", *Science*, 297.5579: 218–22, https://doi.org/10.1126/science.1072708

3. Communication and Knowledge Transfer on Climate Change in the Philippines
The Case of Palawan

Thomas Friedrich

Separately from its physical reality, climate change is a travelling idea (Hulme 2009). Through numerous policies, laws and regulations, the global discourse on climate change affects many people, irrespective of how strongly they experience the consequences of a changing climate. The idea travels from global to local via a long chain of communication and translation. Along the way, knowledge becomes detached from meaning (Jasanoff 2010). This chapter considers the Philippine island of Palawan to show how an idea can be re-integrated into a meaningful context during multiple translations from source to destination in local ontologies. It demonstrates that the local reception of climate-change discourse is influenced by pre-existing systems of knowledge and meaning that are reproduced by circular rather than unidirectional, top-down communication. Irrespective of scientific accuracy, climate change thus becomes a coherent, plausible, and tangible concept regarding what people already know, believe and experience. Using empirical data collected in multi-method fieldwork, this chapter shows that sense-making is a multi-layered process, in which various sources of information play a decisive role in how climate change is comprehended and communicated. Using the example of a lay theatre performance, the chapter demonstrates how the reproduction and dissemination of the local notion of climate change unfolds, and offers recommendations for climate communicators.

https://doi.org/10.11647/OBP.0212.03

Introduction: Translating a Travelling Idea

There is an overwhelming scientific consensus that global warming is occurring as a result of human activity (Cook et al. 2013). Although the diffusion of public climate-change skepticism remains an issue especially in English-speaking countries, publics around the world largely accept that climate change is a cause of concern (Engels et al. 2013). National surveys (e.g. TNS Opinion & Social 2017) indicate that climate science, and how it is conveyed, strongly influence both what is generally known about climate change and public attitudes towards it. Thus, it may be assumed that scientific knowledge content, insights and main conclusions have been successfully communicated throughout the world. In that sense, it has been stated that climate change is not only a measurable physical phenomenon but has also developed into a travelling idea (Hulme 2009).[1] The discourse on climate change has become a prevalent part of information and education campaigns, environmental laws and policy measures. Climate change *as an idea* is affecting many people, regardless of whether they have personally experienced its effects. This idea has travelled in a top-down direction; it was initially developed by (predominantly natural) scientists from different disciplines and countries, represented by the Intergovernmental Panel on Climate Change (IPCC). The global climate-change discourse has informed billions of lay people that our planet's climate is changing due to anthropogenic causes, which will have severe consequences. In regularly published IPCC reports that contain up-to-date scientific knowledge about climate change, the idea has been reproduced and disseminated widely. Subsequently, through a network of media, politics, and other modes of communication, it then engages in a complex process of knowledge transfer and transformation from the global to the local level before it eventually reaches its final recipients—local people with

---

1   Although Hulme never used the term travelling idea, he has described climate change as "an idea that now travels well beyond its origins in the natural sciences. And as this idea meets new cultures on its travels and encounters the worlds of politics, economics, popular culture, commerce and religion—often through the interposing role of the media—climate change takes on new meanings and serves new purposes" (Hulme 2009: xxvi). For a more comprehensive description of the concept of travelling ideas at the global and organizational levels, see Czarniawska-Joerges and Sevón (1996; 2005).

diverse cultural backgrounds and epistemologies who try to make sense of (the idea of) climate change. The research questions underlying this chapter are how this knowledge transfer takes place in a local context, i.e. which are the key players involved, and how the local reception and reproduction of the scientific discourse on climate change can be captured empirically. This chapter outlines this chain of communication and translation of scientific knowledge about climate change from the global to the local level, using the example of the island of Palawan in the southwest of the Philippines.

Of course, this linear model of communication is very simplified. To analyze and fully understand the process of knowledge dissemination, loops and feedbacks must also be taken into consideration (Weingart, Engels, and Pansegrau 2000), especially at the local level. Those who initiated the global discourse (namely climate scientists) are less diverse than the multipliers and recipients of this discourse around the world. The more it trickles down into the manifold realms of politics, society, and culture, the more it interacts with other national discourses, local narratives, and traditional ontologies. Thus, not only knowledge is of utter importance when it comes to climate change communication, but also meaning (Jasanoff 2010). Therefore, the way in which people make sense of climate change relates not only to what they know or do not know about it, but also to how they *understand* the idea, and what it actually means to them when they are told that the climate is changing, that the sea level is rising, or that extreme weather events are more likely to occur in the future. Do they understand the fundamental causalities of the global discourse in the same way as scientists do, or are there alternative models of explanation? And why do some of these models eventually prevail within a society, while others do not? For Jasanoff, the fact that climate science "cuts against the grain of ordinary human experience" (Jasanoff 2010: 237), and thus produces local discordances, relates to how it separates scientific knowledge from meaning on a global level during the process of scientific assessment: "Scientific assessments such as those of the Intergovernmental Panel on Climate Change helped establish climate change as a global phenomenon, but in the process they detached knowledge from meaning. Climate facts arise from impersonal observation whereas meanings emerge from embedded experience" (Jasanoff 2010: 233). In the process of communicating and translating scientific facts about climate change, the climate discourse has encountered various, in part, conflicting local epistemologies and cultures of knowledge, e.g. cultural models of nature that contradict each other (Beck 2007; Bostrom and Lashof 2007; Jasanoff 2010). This raises the question of how exactly (global) scientific knowledge and (local) cultural knowledge intertwine or, in other words, how the "counterintuitive nature of global warming causation" (Rudiak-Gould 2014a: 370) fits into more traditional ethno-ecological beliefs. Indeed, scientific findings are themselves the result of the co-production of knowledge and are thus affected by cultural constraints:

> Scientific knowledge, in particular, is not a transcendent mirror of reality. It both embeds and is embedded in social practices, identities, norms, conventions, discourses, instruments and institutions—in short, in all the building blocks of what we term the *social*. (Jasanoff 2004b: 3, original emphasis)

> [S]cientific facts bearing on the global environment never take root in a neutral interpretive field; they are dropped into contexts that have already been conditioned to produce distinctive cultural responses to scientific claims. (Jasanoff 2010: 240)

So, what does this mean for climate change communication? I use the case of Palawan to illustrate that what local people know about climate change cannot be separated from the meaning they attach to the concept. As Jasanoff has pointed out, meaning derives from experiences that are embedded in specific environmental, social, and cultural contexts. It is therefore reasonable to assume that the way people make sense of climate change is inextricably linked to these contexts. As outlined below, what climate change means to the inhabitants of Palawan strongly depends on their cultural model of nature, i.e., on basic assumptions about their natural environment and the weather, as well as past and everyday experiences, national and local discourses, and narratives that frame the idea and constantly reproduce it in its localized meaning. Hence, the complex interconnections of climate change knowledge with associated knowledge domains must be taken into consideration in order to fully understand how the travelling idea is integrated into pre-existing systems of knowledge and meaning. In this regard, the term *making sense* refers to the process of making foreign knowledge familiar within the context(s) of present beliefs by establishing the best possible coherence and plausibility.

In social representations theory (cf. Smith and Joffe 2013), the process lay people use to make (common) sense of unfamiliar information is described as "anchoring", i.e., the classification and naming of "foreign and threatening phenomena in terms that resonate with those attempting to understand the phenomena" (Smith and Joffe 2013: 18; see Chapter 2, Greenland). As the Palawan case shows, those resonating terms and categories are a crucial element of understanding how local individuals make sense of climate change. Furthermore, the case demonstrates that anchoring also takes place when the English term *climate change* is not translated into the local language but is instead retained and given a new specific meaning. According to social representations theory, sense-making of unfamiliar and uncertain information depends on the interplay between personal everyday experiences and media coverage (Smith and Joffe 2013: 28). However, I argue that more dependencies should be considered. Individuals' images, associations, and representations about climate change are also determined by national or local discourses, which in turn depend in many ways on media coverage and personal experiences (or lack thereof), and also influence them. Together they form the framework that determines how the traveling idea of climate change is eventually received, contextualized, and made familiar locally.

This is especially true for Palawan and many other places in the Global South, where people's access to information about climate change and media access in general is less comprehensive. Therefore, my extended field research on Palawan not only focused on how people receive the discourse on climate change in terms of how they obtain relevant information, but also on how that discourse is locally reproduced—i.e., how climate change knowledge is distributed and communicated apart from unidirectional media consumption.

Palawan is home to a multitude of sources from which knowledge about (and thus public awareness of) climate change is drawn and disseminated. For example, media sources (radio, TV, internet, and, to a lesser degree, newspapers) are quite accessible on the island, especially in the capital Puerto Princesa City. Furthermore, the people of Palawan learn about climate change in educational institutions (governmental as well as non-governmental), discuss it in community meetings, hear about it in churches, are affected by its political implications and implementations, and talk about it with family, friends, and neighbors. Scientific terms such as *climate change, global warming, greenhouse effect,* or *sea level rise* are introduced to society through these various information channels. In order to examine how climate-change discourse is received and locally reproduced, one of my fieldwork goals was to identify relevant "translation regimes" (De Wit 2015), by which I mean all local agents, institutions, communication patterns, discourses, and narratives that help to translate the global discourse into a generally intelligible common sense understanding, and thus significantly shape the perception, conceptualization, and communication about climate change. Translation regimes—which include, but are not limited to, the media—embed the scientific notion of anthropogenic climate change into meaningful local contexts. Together they provide the socio-cultural, discursive, and epistemic basis for how knowledge about climate change is distributed, organized, and structured throughout society. Since local translation regimes are integral parts of local society and culture, they enable us to research processes of anchoring, translating, interpreting, reproducing, and thus sense-making of climate change as dialectical rather than unidirectional. Below I describe one such regime from Palawan—an amateur theatre—to demonstrate how local concepts, narratives, and values pertaining to the relationship between the island's population and their natural environment are reflected in climate change communication. In the conclusion, I recommend that climate change communicators pay more attention to local peculiarities and corresponding translation regimes in order to understand how people make sense of climate change, and why they do it in a particular way.

Localizing the Study of Climate Communication

Much of the research to date on public perceptions of climate change has been based on large-scale national and international surveys (e.g. TNS Opinion & Social 2017). There have been few qualitative investigations of how people integrate scientific knowledge into existing local knowledge and meaning contexts. Ethnographic fieldwork is particularly suited to this purpose. Qualitative methods not only help to determine what people know and how their knowledge is embedded into broader socio-cultural contexts; they can also elicit tacit forms of knowledge that people rarely articulate explicitly, if at all, e.g., in interviews. This chapter is based on ethnographic fieldwork that used multiple methods of collecting and analyzing empirical data to obtain a more holistic view of how local knowledge about climate change is reproduced, distributed, and applied discursively. As further described in the methods section, these include participant observation, guided interviews, a survey, and experimental methods of cognitive anthropology.

One of my main findings was that local investigations of climate change knowledge, and the discourses and narratives that shape communication about it, should examine local ecological and weather knowledge that strongly influences people's experiences with their environment, and thus affects how global climate-change discourse is received locally. As Strauss, for example, has shown in her ethnographic description of the Foehn in the Swiss Alps, the people of the small town of Leukerbad experience and feel this warm and dry wind, which has become as much a vital part of their identity as the surrounding mountains. "Wind" Strauss states "is part of the landscape" (Strauss 2007: 179). Common ecological knowledge in general, as well as basic knowledge about the weather, is locally anchored and, to varying degrees, socially shared. It is thus not merely information or factual knowledge. It also includes tacit knowledge, embodied knowledge, basic assumptions, norms and values, as much as it is expressed in feelings, discourses, and social behavior. This kind of local knowledge is generated and reproduced based on everyday experiences of—and interactions with—(changing) environments, e.g., by the regular use of natural resources or weather experiences. Yet, since it is also the subject of constant negotiations, it is a dynamic form of knowledge. Ingold (2010) employs a phenomenological approach to illustrate how knowledge is acquired by physically moving through the "weather-world". This strongly resembles the idea of a "local epistemology" (Friedrich 2018), but lacks the social dimension of knowledge, i.e., how it is shared and socially construed. On the scale of national states, this is what Jasanoff emphasizes in her concept of "civic epistemology", by which she means all "institutionalized practices by which members of a given society test and deploy knowledge claims used as a basis for making collective choices" (Jasanoff 2007: 255). This does not mean, however, that recognition of civic or local epistemologies implies a dismissal of scientific standards, or that local knowledge should be romanticized. Rather, that acknowledging different epistemologies, i.e., processes of knowledge co- and reproduction, allows for a deeper understanding of why local translation and integration of global knowledge varies widely (Lutes 1998; Roncoli, Crane, and Orlove 2009).

The difficulty of reconciling (global) scientific and (local) non-scientific systems of knowledge is demonstrated by exploring the intersection of global and local knowledge about climate change. It has been pointed out many times that people do not experience "the climate" as a statistical figure, but the weather, its variations, and environmental changes (West and Vásquez-León 2003; Bostrom and Lashof 2007; Peterson and Broad 2009; Rudiak-Gould 2012, 2013b). Unlike knowledge about natural environments and the weather, natural scientific knowledge about climate change is impersonal, apolitical, intangible, and universal. Global climate change takes place on a huge scale, both spatially and temporally; the physical causes and effects are located far from each other. At the same time, the climate is a slowly changing system. Climate scientists assert that many climatic changes that can be observed today are the result of anthropogenic greenhouse gas emissions from past decades, which means that the benefits of recently implemented measures to reduce carbon emissions will not be realized for decades later, if at all. From a scientific point of view, a direct link between the behavior of one individual and climatic or weather variations cannot be established. However, to make sense of perceived changes in weather and the environment, people do make this link to find plausible explanations based on their ontologies. For example, Roncoli et al. (2003) found that farmers' interpretations of scientific weather forecasts in Northern Burkina Faso are strongly influenced by personal interests and their concept of rain. The authors conclude that: "[S]cientific and technical knowledge is not a 'product' that can be pre-packaged and delivered to 'users' without its being altered by its incorporation into a different set of meanings and relations from those that produced such knowledge" (Roncoli et al. 2003: 197). Therefore, from an anthropological point of view, local heuristics or explanatory models are of the utmost importance. Instead of seeing them as simple misinterpretations of scientific concepts that have to be corrected (c.f. Bostrom and Lashof 2007; Chen 2011), they can also be understood as a cultural expression of confirmation bias, i.e., the psychological phenomenon that people accept new information only when it is consistent with their pre-existing knowledge, and reject it when it challenges ingrained concepts (Rayner 2003). This applies to environmental or pollution concerns as much as climate change, where political, moral, and conceptual plausibility has been demonstrated to be more important than physical causality (Douglas and Wildavsky 1982; Douglas 1992; Jasanoff 2010; Rudiak-Gould 2014b). Thus, anthropologists are not interested in whether local people have an "accurate" understanding of the complex notion of scientific climate change. They instead ask how people conceptualize the idea, how they relate it to their environment, how their alternative explanatory models fit with what they already know and believe, which local translation regimes influence the way they conceptualize and talk about climate change, and whether or not their cultural models are challenged by science or other "potentially incongruent sources and knowledge practices" (Hastrup 2015: 142).

Models of Knowledge Transfer

This context relates to associated fields like translation, (risk) communication and public understanding of science (Rudiak-Gould 2012). One of the shortcomings in past research on science communication has been identified as the expert/lay dichotomy that distinguishes between two groups believed to be homogeneous, i.e., "scientists", as the discourse producers, and "the public", as the discourse recipients (Thompson and Rayner 1998). The underlying assumption of this dichotomy, "that it is possible to separate what the risks really are from what the public erroneously believe them to be" (Thompson and Rayner 1998: 165), not only simplifies the complex discourse and ignores the heterogeneity of those that constitute it. It has also led to incorrect conclusions in social scientific research, public information campaigns, and policy-making. With reference to the pioneering work of Kempton et al. (1995), who showed among other things how fundamentally

US-Americans' perceptions of climate change are based on different pre-existing cultural models that are widely shared, Thompson and Rayner conclude that "the accuracy of the detailed information that members of the public are receiving from media discussion or public awareness campaigns is largely irrelevant" (Thompson and Rayner 1998,: 148). Thus, they conclude that more precise science communication does not necessarily lead to a more accurate public understanding of the problem. For example, a number of studies have shown a widespread phenomenon, which I also encountered on Palawan (described in more detail below), that lay people usually confuse climate change with ozone depletion (Löfstedt 1991; Kempton 1991; Bostrom et al. 1994; Kempton, Boster, and Hartley 1995; Thompson and Rayner 1998; Smith and Joffe 2013; Friedrich 2017). A popular "solution" for such phenomena would be to correct this misconception by better communicating scientific knowledge, also known as the *(information) deficit model*. By providing targeted education about the actual causal interconnections, it is assumed that the public will eventually see the world the way scientists do. This unidirectional sender/receiver structure of communication has been criticized as psychologically and sociologically naïve (Stern 1992; Kearney 1994; Wynne 1995; Thompson and Rayner 1998; Blake 1999; Weingart, Engels, and Pansegrau 2000; Jasanoff 2007; Hulme 2009; Jasanoff 2010; Rudiak-Gould 2013b). Indeed, the deficit model ignores the fact that the acquisition of new knowledge is a selective process, and that receiving new information does not necessarily lead to the understanding intended. In order to integrate it into pre-existing systems of knowledge and meaning, there must be a certain amount of coherence and plausibility, since "a series of isolated facts or details do not create meaning" (Kearney 1994: 430). Creating meaning, however, is a creative and thus active process that does not fit into a unidirectional structure of communication. Instead of trying to correct "wrong" concepts with more detailed scientific information, these concepts should be taken into serious consideration. Another limitation of the deficit model is what environmental psychology labels the knowledge-behavior or value-action gap. This refers to the problem that people apparently do not act as desired, for example by changing their consumption patterns, even though they seem to know enough about their individual and collective impact on the environment, and hold high environmental values (Blake 1999).

In comparison, the *sociocultural model* explains the relationship between knowledge and behavior much better; it is less knowledge-based and considers norms, values and social actors. Because, as Jasanoff has stated, "without human actors [...] even scientific claims have no power to move others" (Jasanoff 2004a: 36). Among others, she has pointed out that science alone cannot provide a moral basis for action (Milton 1996, p. 124; Thompson and Rayner 1998; Jasanoff 2007). Instead of a unidirectional transfer of climate change knowledge, the sociocultural model assumes a more dialogical communication along "cultural circuits" (Hulme 2009), like science, politics, media, and the public that represent distinct domains in which "[m]essages about climate change have no starting point and no ending point; they travel around this circuitry, changing frame, form and meaning as they go" (Hulme 2009: 221).

A third model that more strongly emphasizes the meaning of discourses and their role in the selectivity of knowledge acquisition can be called the *discourse model*. Discourses provide an epistemological and normative structure to determine what particular information is relevant, and what can be ignored or rejected. By considering the national and local translation regimes—i.e., social agents, institutions, communicational patterns, narratives, and discourses that mediate and translate the global discourse on climate change—the manifold local manifestations of this discourse eventually become explainable. Translation regimes show us how, where, and by whom relevant knowledge is actively shared, verbally or non-verbally. Identifying and analyzing such regimes is indispensable to delineate how the scientific discourse is being received and reproduced on a local scale.

Language is another significant factor that permeates all the models described. The anthropologist Rudiak-Gould emphasized that the communication of scientific climate change knowledge is not only an issue in terms of translating ideas, concepts and cultural models; it is also a linguistic problem:

> Climate change communication is ultimately an issue of translation: the cultural translation from scholarly communities to citizens; the cultural translation from Western and other elite developers of climate science to

indigenous people and other non-Westerners; the linguistic translation from specialized climatological jargon to the colloquial language of citizens; and the linguistic translation from English, and other languages in which the notion of anthropogenic global warming has been formulated and studied, to the languages of those who are called upon to prevent or prepare for it. (Rudiak-Gould 2012: 46)

During his fieldwork on the Marshall Islands, Rudiak-Gould found that the local term that is used to translate climate change into Marshallese language already holds various meanings, and thus influences how climate change as a concept is understood. He assumes that this kind of mistranslation (deficit model) or reinterpretation (sociocultural and discourse model) occurs in many other societies and languages as well. Instead of dismissing those local concepts as a result of failed communication, however, he suggests that they may provide the opportunity for real dialogue (Rudiak-Gould 2012, 2014b). In this context, anthropologists, linguists and other social scientists familiar with local patterns of meaning can make a decisive contribution to improving climate communication.

In the Philippines, the English term *climate change* is predominately used in public discourse, since English is the country's second lingua franca. For instance, in national television broadcasts, where "Taglish" is predominately spoken (a common mix of the first lingua franca, Tagalog or Filipino, and English), the term *climate change* is clearly preferred to its Tagalog translation *pagbabago ng klima*. However, even if the same term is used as in global scientific discourse, this does not mean that its meaning has not changed or will not change further along the chain of communication. The question of translation does not disappear just because there is no linguistic translation. What remains is cultural translation. The Philippine example therefore demonstrates why a distinction must be made in climate change communication between knowledge and meaning as much as between local and global discourses.

Communicating Climate Change in the Philippines[2]

As an archipelago, the Philippines is among the countries most vulnerable to climate change. The IPCC defines climate change vulnerability as "the degree to which geophysical, biological and socio-economic systems are susceptible to, and unable to cope with, adverse impacts of climate change" (Intergovernmental Panel on Climate Change 2007: 21). In international comparisons, the Philippines is always ranked highest because even without climate change, it experiences multiple extreme weather events that regularly cause massive damage and casualties. Each year between May and November, around twenty tropical cyclones enter the Philippine Area of Responsibility, almost half of which make landfall. According to the Global Climate Risk Index (Eckstein, Künzel, and Schäfer 2017), the Philippines is ranked the fifth most affected by weather extremes and subsequent events like floods or mudslides between 1997 and 2016. In 2013, it even led the ranking as a consequence of super typhoon Yolanda (international name: Haiyan) that is considered to be the strongest typhoon that has ever made landfall (Kreft et al. 2014). Against this background, it becomes comprehensible that numerous Philippine policies, laws, and regulations have incorporated natural and environmental hazards and their consequences. For example, a natural disaster discourse is virtually ubiquitous in the country. As elaborated below, this discourse plays a pivotal role in the way climate change is commonly understood and adopted in policy-making processes. On Palawan, the dovetail of natural disaster discourse and climate-change discourse is an integral part of local translation regimes that help to translate the scientific discourse on climate change into a more intelligible and tangible understanding, taking actual experiences (or lack thereof) into account. In order to understand climate change communication on Palawan, it is thus worthwhile to first examine official national climate policy and its implications for local contexts.

The Philippines was one of the first countries in the world to make climate change a legal issue. As early as 1991, then-President Corazon Aquino signed Administrative Order No. 220, which provided for the

---

2   Modified versions of this and the following section can also be found in Friedrich (2018).

foundation of a national climate change committee. The first IPCC assessment report was published just one year before the Philippines officially accepted the "mounting scientific evidence of an impending global warming" (President of the Philippines 1991). In 1994, it ratified the United Nations Framework Convention on Climate Change and nine years later, the Kyoto Protocol. In 2007, the fourth IPCC assessment report was published, which again had a strong impact on Philippine legislation. President Gloria Macapagal-Arroyo issued a decree that year which stated, with regard to the IPCC, that "climate change poses serious threats to the lives and welfare of the people, especially the poor households, and sustainable development of the country" (President of the Philippines 2007). The decree also ordered the foundation of a task force on climate change that eventually merged with the former climate change committee in 2009 to form the current Climate Change Commission (Republic of the Philippines 2009). As demonstrated here, the Philippines is understood as a country that is very vulnerable to the impacts of climate change. It is located within the so-called Typhoon Belt, "experiencing [an] unusual number of high-intensity typhoons that have wrought devastations and anguished to our people" (President of the Philippines 2007). The commission was tasked with informing and educating the public about the issue of climate change, raising awareness of its adverse effects, and mobilizing appropriate responses. The dissemination of climate change knowledge therefore became official, and eventually resulted in a country-specific discourse that explicitly prioritizes knowledge "from the Philippine perspective", and adaptation as a preferred strategy to deal with climate change (Climate Change Commission 2011). According to the National Climate Change Action Plan, the communication of climate change knowledge must consider the specific local context:

> Having access to relevant information and localizing it from the Philippine perspective: There is a lot of scientific information about climate change in [sic] the global level. [...] Climate-change impacts vary from one place to another and so researches [sic] on the local impacts are important. [...] Development of [...] communication materials should consider who the target is and what type of materials are suitable to them. (Climate Change Commission 2011: 33)

What is locally known about climate change in the Philippines, therefore, is not just the result of an incidental loss of information along a long chain of communicating, translating, and simplifying the complex and comprehensive scientific knowledge from the global to the local level (as in the children's game "Chinese whispers"). Instead, the Philippine government intentionally pre-selects and stresses particular knowledge content and discards others. Considering the country's continuous experience of extreme weather events and natural hazards, and its overall educational situation, this policy reflects a very pragmatic approach to the climate change issue, and therefore serves as a good example of how political discourses as part of laws and regulations, initiatives and programmes, but also across the media, continuously constitute access to, dissemination of, and interpretation of knowledge. This also applies to the local scale, where global scientific and national political discourses encounter specific local discourses and narratives that may or may not be compatible with one another. In the following this is shown by the example of the island of Palawan, which is special on many levels compared to the rest of the archipelago.

"Palawan is Different!"

Consistent with the Filipino population in general (Social Weather Stations 2013), the vast majority of the people that I encountered during my fieldwork on Palawan were convinced that climate change is real: fifty out of fifty-three respondents (94%) to my survey[3] agreed with the statement "Climate change is a calamity which is happening in the Philippines right now". However, 84% also agreed with the statement "We experience climate change in Palawan. But other parts of the Philippines experience it much stronger than us". This is just one of many indicators that a notably large percentage of my interlocutors continually made a strong distinction between their island and the rest of the country. This predominately positive demarcation was also expressed in terms of safety issues, general quality of life, environmental matters and even geophysical phenomena. Respondents stated that,

---

3   For a detailed description of how the data were collected and how the sample was constructed, see subsequent section.

unlike in the Philippines in general, there are no earthquakes, volcanoes or very strong typhoons on Palawan. This claim is largely supported by scientific accounts, since the island does have some exceptional features regarding its tectonic structure. Although, like most of the country, it is geologically part of the Sunda Plate, the island is not affected by orogenic movements (i.e., the formation of mountains). As a significant amount of endemic species of flora and fauna, as well as different soil conditions indicate, Palawan is more closely related to Borneo and the continental part of South East Asia as opposed to the remaining archipelago, which is of volcanic origin (Baillie, Evangelista, and Inciong 2000; Esselstyn, Widmann, and Heaney 2004). This also means that Palawan is not part of the Pacific Ring of Fire and is thus considered to be an "aseismic region" (Lagmay et al. 2009). Moreover, it does not lie within the Typhoon Belt, which makes the island less vulnerable to the strong typhoons that usually approach the country from the Eastern Pacific. However, as a deeper look into historical typhoon paths reveals and the typhoon season of 2017 has once again demonstrated, the island is certainly not typhoon-free (Palawan Council for Sustainable Development 2004: 49; WWF Philippines and BPI Foundation 2014).

Another distinctive feature that is always brought up by Palawan's inhabitants when asked what is so special about their island is its unique, albeit vulnerable, natural environment. The island is known as "the last (ecological) frontier" in the Philippines due to its massive forest cover,[4] biodiversity, healthy coral reefs and low population density. Almost 40% of Philippine fauna can be found here, including many endemic and endangered species (Palawan Council for Sustainable Development 2004). Palawan is also home to about one-fourth of the Philippines' intact mangrove forests (Long and Giri 2011). Despite being the largest province in the Philippines by area, only around 1% of the total population lives there. In 1990, the UNESCO declared the province a Man and Biosphere Reserve and a model region for sustainable development (United Nations Educational, Scientific, and Cultural Organization 2011). Two years later, a national law was passed, known as the Strategic Environmental Plan for Palawan Act, which led to the establishment of

---

4   Its forest cover of 46.5% is considerably higher than the Philippine average of 22.8% (calculation based on Food and Agriculture Organization of the United Nations (2010); Forest Management Bureau (2015)).

the Palawan Council for Sustainable Development (PCSD) (Republic of the Philippines 1992). For more than twenty-five years, the PCSD has been responsible for extensive programmes to protect, conserve and develop the island's natural resources, and to educate its population on climate issues. It has also become a regional translation regime that impacts how the discourse on climate change is perceived on the island.

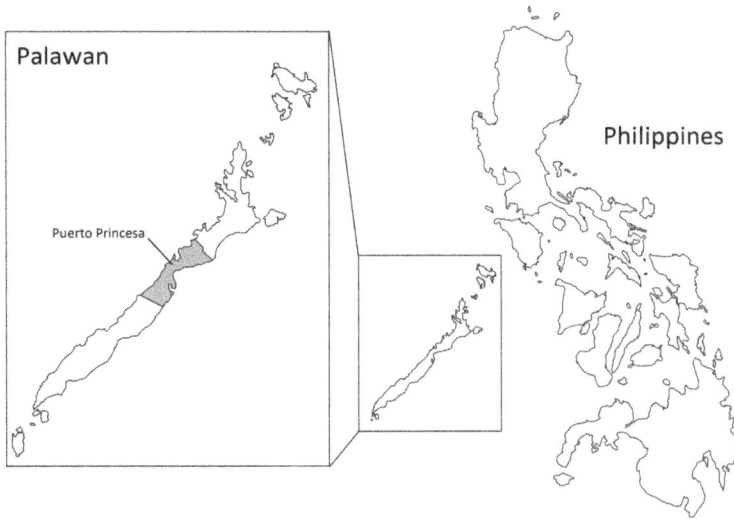

Fig. 3.1 The island of Palawan with its capital region that consists of both urban and rural areas (source: author, 2020).

On a local scale, the same can be said for the city administration of Palawan's capital, Puerto Princesa City. My fieldwork and all my data collection took place within its municipal boundaries. In the same year the PCSD was created, 1992, a new mayor took office in Puerto Princesa City, who was praised shortly afterwards by the United Nations Environment Programme in 1997 as "the first Filipino political leader to make environmental protection the centerpiece of his administration" (UNEP 2004: 2011). At this time, and for the third time in a row, the city won the national contest for the "Cleanest and Greenest Component City in the Philippines" due to its very successful "operation cleanliness" (Tagalog: *oplan linis*). During this mayor's term of office until 2013 (with a short break between 2001 and 2002), Puerto Princesa City rapidly developed from a former penal colony administration into a best practice example for successful climate change adaptation, and one of South Asia's prime eco-tourism destinations (Department of Environment and Natural Resources 2012; WWF Philippines and BPI Foundation 2014). As my overall findings indicate, those and other policy implementations have had a huge and sustainable impact on the thinking and behavior of the city's inhabitants regarding how they value nature and how they understand their individual and collective impacts on natural processes. In a nutshell, environmentalism virtually became common sense.

One of the key components of local environmental rhetoric has been about not only protecting but also improving the island's vegetation by planting trees to enhance its resilience to weather-related disasters and sea level rise. Palawan hosts two of Southeast Asia's oldest and most popular tree planting festivals (Friedrich 2017). Bearing in mind the specific historical, political, and environmental context of Palawan and its capital, an intriguing finding of my research thus becomes clear—that the island's above-mentioned biogeographical exceptionalities strongly correlate with people's attitudes and behavior towards their natural environment and climate change. As my data shows, there is a strong consensual perception of climate change as just one hazard among others. Furthermore, an overwhelming majority of my sample reported that what they understood as "proper environmental behavior" was also understood as a significant factor in preventing natural disasters on Palawan, including climate change.

Methods

Determining how the people of Palawan understand and make sense of climate change entails intricate methodological challenges. While there are many ways to assess what people know about a certain subject, it is much more difficult to determine how they comprehend a concept. You might ask them what they know about climate change and they might tell you a lot about it (irrespective of whether it is scientifically correct), but evaluating *understanding* requires much more than factual knowledge. Although people might understand climate change somehow, they may not be able to verbalize it explicitly. As social representations theory states, anchoring and comprehending an unfamiliar scientific concept like climate change requires a system of reference that constitutes whether or not new knowledge content makes sense (cf. Smith and Joffe 2013: 18). From a local perspective of understanding, this function is commonly fulfilled by pre-existing systems of knowledge and meaning that include basic, underlying assumptions. In the case of climate change, these assumptions include fundamental beliefs about how nature and the weather generally *work*, i.e., how natural processes are interrelated or what impacts humans have on these processes and vice versa. This kind of tacit knowledge can be shared widely within a social group. However, possession of such cultural models does not necessarily enable its members to reflect them. Despite this knowledge being taken for granted—or perhaps because of that—people do not talk about it. Therefore, in order to reconstruct local people's perspectives on climate change, I used a strongly bottom-up, data-driven approach with multiple methods of data collection that also took implicit forms of knowledge into account and allowed for maximum empirical openness by reducing the probability of biasing the results with my own underlying assumptions about the scientific concept of global climate change. Following this methodological approach, which was largely inspired by Grounded Theory (Glaser and Strauss 1967), each method of data collection was built on previous collections and continuously compared. This section briefly presents each of the data collection methods as well as some of their crucial results.[5] Subsequently, more emphasis will be put on the participant observation part of my fieldwork, with a more practical example of how climate change knowledge is communicated, transferred, and disseminated on Palawan beyond media consumption.

My fieldwork took place for a total of seven months between 2013 and 2015. The central data collection was carried out between October 2013 and March 2014. During that time, I successively collected various data formats from almost 100 people living in the municipality of Puerto Princesa City, using a multi-method approach that included qualitative as well as quantitative methods. In addition to participant observation,

---

5   For a detailed description, reflection, and discussion of the overall methodology, see Friedrich (2017).

conducting interviews and administering a survey, further methods from cognitive anthropology have been applied, namely freelists and pilesorts, which proved particularly suitable to my research question.[6] Each of these data formats, collected in different, partly overlapping groups of informants, represented a certain perspective on the topic; however, they complemented each other and eventually provided a more complete picture. Accordingly, my overall findings are the result of the triangulation of *all* these methods.

Data analysis was carried out at the same time as data collection and thus also followed the methodological approach of Grounded Theory, leading to the emergence and development of empirically based, local concepts, categories, and ideas, which in turn led to hypotheses that were integrated into the design of the next phase of data collection. For instance, most of the survey statements with which participants were asked to agree or disagree were the result of a cluster analysis of the pilesorts that were mainly based on the results of the freelists. Many of those statements would never have found their way into the survey if I had designed it at the beginning of my fieldwork. "Being there" and collecting data was an essential requirement for eliciting what eventually proved to be *culturally relevant* statements and interpreting the overall data adequately. By constantly cross-referencing data formats with background information and personal observations, this exploratory mix of methods produced findings that were more valid and plausible within their sociocultural, historical, and political context. Furthermore, participant observation was particularly helpful for comprehending how people's knowledge about climate change was put into practice in everyday life and on special occasions.

---

6   Freelists and pilesorts are techniques used to elicit items, elements or members of a certain cultural domain. They are lists of free associations that help to detect what cognitive anthropologists describe as cultural domains or mental categories, i.e., culturally relevant clusters of items or subdomains (e.g., "edible fishes"). For an overview of how those and other cognitive methods are applied, see Borgatti and Halgin (2013). For a more detailed description of how I conducted and analyzed the freelists and pilesorts, see Friedrich (2018).

Findings

One of the freelist tasks that was administered in both English and Tagalog was to note down personal associations with the term *climate change*. The results show that more than half of the respondents who performed this task (n = 31) linked climate change to natural disasters and extreme weather events such as those occurring in the Philippines. Almost half of them mentioned predominantly negative effects on society and the general weather. The following interview sequence with Abraham[7] provides an example:

> Thomas: Personally, have you already experienced the impacts of climate change?
>
> Abraham: I'm sick because of climate change.
>
> Thomas: And why is that?
>
> Abraham: You know, the changing weather. Sometimes I wonder, is it Christmas today? You know? Sometimes we experience this [changes] in November or October, but now even early morning. Not only today but also in the past. (Interview with Abraham, 17 January 2014)

By "Christmas" Abraham refers to the northeast monsoon called *amihan* that brings cool winds to Palawan between October/November and February/March. The beginning of this dry season more or less coincides with the Christmas season that starts around September in the Philippines. Abraham wondered about the perceived shift in seasons that was expressed by many of my interlocutors, including too much or too little precipitation or somehow "wrong" temperatures in the past few years. Tetina, a woman from one of the Indigenous groups of Palawan, the Tagbanwa, gives another example:

> Thomas: How would you describe, in general, the weather here in Puerto Princesa City or Palawan?
>
> Tetina: With regards to weather my observations before was (sic): If it's summertime, it is summertime—it is the dry season. And if it is the rainy season, it is raining. But now, I think, like three years ago, there are changes based on my personal observations. (Interview with Tetina, 21 January 2014)

---

7   To maintain anonymity, an alias was created for all informants.

Respondents provided a rather diverse set of answers in the freelists. However, when specifically asked about the causes and effects of climate change, about two-thirds named general harm to the environment as a major cause. For almost half of them this included damage caused to trees or the forest in general. For about one-third the wrong use of waste, e.g., burning plastic, was another culprit. In sum, over 80% of the causes mentioned could be classified as human induced, which is also reflected more explicitly in the survey statement "Global warming is man-made", which 87% of participants (n = 53) agreed with. Climate change skepticism does not appear to be an issue on Palawan. Apart from natural disasters and extreme weather events, climate change effects were strongly associated with adverse effects on the environment, general weather, and society.

The freelists results were subsequently incorporated into a pilesort (n = 34), i.e., a card-sorting task. The most common terms mentioned in all freelists were identified and written on thirty-two cards, combined with further terms, e.g., from the scientific discourse on climate change, and then randomly shuffled and handed over to another set of participants who independently sorted them into piles according to what they believed to be affiliation. At the end of each sorting activity, participants were asked to explain the topic of each of their piles. The answers given to this question were of the utmost importance for understanding the categories or domains of thinking used to cognitively structure those terms within a common knowledge context. A cluster analysis of all the pilesorts revealed a dominant pattern of domains that clearly distinguished between what were perceived to be positive and negative human activities (e.g., tree planting and burning garbage, respectively) and their impacts on nature (see Friedrich 2018 for an analysis).

A consensus analysis of the pilesorts and the survey showed an overwhelming agreement regarding how my interlocutors thought about climate change and their natural environment. Climaxing in several survey statements with an agreement up to 100% of all independently interviewed participants, this consensus was so strong that it would not be unreasonable to suggest that it not only applies to my sample but to the population of Puerto Princesa City, or maybe even Palawan in general. For example, one of the statements with 100% agreement was "It's not enough to let nature just recover itself. We have to protect, conserve and restore it actively, for instance by planting trees". Another was "Development means that we have to take good care of our environment. This is the reason why all the tourists come to Palawan". The manifold levels of meaning of those statements are hard to decipher without the necessary background information that was briefly outlined earlier. The importance of collective tree planting activities or the strongly positive connotation of *development* (Tagalog: *unlad*), which expresses itself in tree planting in the sense of *environmental development*, can only be understood by taking Palawan's special features into account (i.e., its biogeographical exceptionalities and the strong environmentalism of its inhabitants, especially in the capital). As my data show, people strongly correlate both features and construe a causal relationship between the fact that the island is not affected by earthquakes and rarely hit by strong typhoons, and their environmental behavior. Up to 81% agreed with the statement "Here in Palawan we don't have earthquakes and very strong typhoons like in other parts of the Philippines, because we take much better care of our environment". Accordingly, this fundamental and consensual belief that people's activities have an important impact on natural processes at the local level also frames how climate change is perceived and conceptualized. The term *climate change* was written on one of the pilesort cards and was predominantly sorted in a way that made it a central component of one the main domains of thinking, comprising all other terms that were understood to be natural disasters (Friedrich 2018). This means that climate change is understood as just another natural force that the Philippines experiences every year. In accordance with people's perceptions and moral sense, the idea of climate change is cognitively integrated into a pre-existing system of categories, where it is anchored to make sense locally.

In addition to the methods presented thus far, participant observation was a major source of valuable information during my fieldwork. In several practical examples that included expected as well as unexpected events and happenings, I explored how climate change knowledge is communicated, transferred, and disseminated discursively and in social practices, and triangulated it with my other modes of data collection. The remainder of the chapter focuses on the conditions and circumstances that maintain and affirm the way of thinking described above through practices of knowledge transfer.

Local Knowledge Transfer

A practical case example that I would like to use in the following to illustrate a specific form of conveying knowledge about climate change is that of the Palawan Conservation Corps (PCC), a non-governmental organization located in Puerto Princesa City. The PCC educates and trains young people in the field of environmental protection, value formation, and restoration. The PCC's objective is to improve the future prospects of adolescents from mainly rural areas (what they refer to as "marginalized youth") by teaching them a combination of practical and social skills and environmentally relevant topics. The six-month full-time training programme takes place in a remote campsite. An essential component of the training is rehearsing the so-called ecological theatre caravan (eco caravan). This project consists of several stage plays that were written and developed by some of the organization's staff members, dealing with ecology, species protection, biodiversity, sustainability, conservation of terrestrial and maritime environment, as well as global warming and climate change. As one of PCC's responsible persons told me, the eco caravan is designed to teach "basic ecology" and raise awareness of "the web of life", according to the organization's holistic mission statement that "everything on this planet is interrelated— everything is connected to everything else". Three of the stage plays that explicitly deal with climate change and its consequences are presented in more detail below. At the end of each six-month training course, the PCC trainees performed the plays in primary and secondary schools. Since climate change has not yet been mainstreamed in formal school curricula on Palawan, this represented an introduction to the topic for many of the children in the audience. The dual transfer of information associated with the eco caravan performance made it particularly interesting for my research: PCC staff members teach the adolescent programme participants, who then serve as mediators of this knowledge and transmit it through their performances. In this way, the PCC and its eco caravan serves as an insightful example for a translation regime on Palawan. Since its foundation in 1999 and the implementation of the eco caravan in 2008, the PCC has served as a successful multiplier in the field of ecological and climate change communication, having reached hundreds of individuals.

My first visit to the PCC's training camp was in November 2013, only a few weeks after super typhoon Yolanda devastated large parts of the Philippines (Friedrich 2018). It was located in a forest outside the city center, where the organization had a long-term lease agreement with the city government that included a teaching building and a simple dormitory. During that time, around twenty trainees and two staff members were spending their last month there. When I arrived with one of my key informants, Jesus, I was very optimistic that I would be able to observe what kind of knowledge is communicated, i.e., how a lesson is structured, what sources are used to inform and educate the attendees, and who exactly passes on which information. My original intention was to investigate how the PCC functions as a translation regime. I was hoping to conduct interviews or collect other forms of data in a subsequent visit to assess what the participants actually know about climate change. However, things soon turned out very differently since Jesus, without my knowledge, had announced our visit in advance. People were not only expecting us when we arrived but suddenly and willingly stopped their daily routines to welcome us. As a result, largely unnoticed observations were no longer an option. Instead of being the invisible observer that I wanted to be, I became the center of attention. And as if that wasn't bad enough already, Jesus even introduced me as a climate change specialist from Germany who was there to tell them why our earth is warming and what we can do about it. Unfortunately, this introduced a methodological and ethical dilemma since my talking about climate change would bias the audience and jeopardize future data collection. Obviously, providing information in advance alters people's knowledge—the very knowledge that I was interested in. Contrary to my intention, I was no longer able to uphold the role of an ethnographer, who aims to elicit information from the people under investigation. Instead, I was forced into the role of an educator who was expected to share information. That I was not able to do this and thus establish reciprocity also bothered me from an ethical point of view. Eventually but reluctantly, I accepted my imposed role in this most uncomfortable situation, and delivered a short lecture about the greenhouse effect, melting glaciers, diminishing polar ice shields, rising sea levels, and increasing weather extremes.

The Dilemma of Climate Change Communication

At the end of my talk, to make matters even worse, Jesus opened a question and answer session. Most of the questions related to whether we, in Germany, also have climate change, typhoons like Yolanda, and sea level rise. The most challenging question was the last, from a very interested young man of eighteen years. He wanted to know what the people of Palawan could do to stop climate change. So there I stood, surrounded by potential future environmentalists who are taught how their behavior can positively and negatively impact their natural environment to make their island a cleaner, healthier, and safer place. I realized that my answer would affect the motivation of these young and eager-to-learn people. How could I tell them that whatever they do—no matter how ambitious it may be—will never stop climate change? Why should I take away their hope, make them aware of their powerlessness and thus discourage them by telling them that the Philippine contribution to global greenhouse gas emissions is insignificant compared to the emissions of industrialized and emerging countries, yet they are disproportionally facing the negative effects of a constantly warming planet? This unjust distribution of the causes and effects of climate change particularly applies to Palawan, which is considered "carbon negative" due to its large forest cover and its virtually non-existent heavy industry sector, meaning that it absorbs much more CO_2 than it emits (Department of Environment and Natural Resources 2012). What climate change mitigation measures could I name that did not devalue those young men and women as responsible agents within a perceived "web of life" in which "everything is connected to everything else", including themselves? Which mitigation measures make sense anyway, in a place like this? Using a bicycle rather than a car? The people in front of me had neither. Eating less meat? The small portions they were offered one or two days a week were mainly poultry or pork from the region. This situation highlighted the need for climate change communication and action to adapt to the place and people it addresses in order to be effective. On this occasion, I eventually decided to answer that young man's question with an activity that is already very popular on Palawan—tree planting—and explaining to him how afforestation helps both to mitigate global climate change and, in the case of mangrove planting along the coast, is an effective protective measure against sea level rise.

Fig. 3.2 Giving an unexpected lecture to trainees at PCC's training camp (photo by author, October 2013), CC BY-4.0.

After the question and answer session I sat down, and Jesus briefly summarized my lecture again in Tagalog, while I was critically reviewing it in my mind. Did I describe the complex relationships of climate change correctly, or did I present its relevant processes too simply? Did I forget something important or was there anything that I should have dropped? My new and imposed role as an educator started to take effect. I was wondering how exactly climate change knowledge should be communicated in such a context. Despite my own self-imposed methodological restrictions regarding what I should or shouldn't say, there were obviously good didactical reasons not to use the scientific language of the IPCC with this audience. But how is it possible to translate and facilitate scientific knowledge contexts without altering them in ways that cause misconceptions? The experience of being an educator against my will helped me to truly understand the dilemma of climate change communication, i.e., how to make expert knowledge comprehensible for lay people, without reducing its complex causalities to a degree that enables scientifically untenable heuristics. It became obvious to me that translating scientific key terms into easy-to-understand local terms is not enough, especially when those terms already have manifold connotations. If climate change communicators want to avoid attaching non-scientific meaning to concepts, interconnections, and causalities of scientific climate-change discourse, they need to ensure that knowledge and meaning are transferred together. But how is simplified communication about climate change then even possible? Sitting on a plastic chair in a remote forest camp on a Philippine island, I perceived myself to be at a possible end of the imaginary line of communication along the scientific discourse on climate change from its global production to its local reception. Even though I was unable to collect the data that I originally intended, the unexpected role play highlighted the difficulty (or even impossibility) of communicating climate change knowledge that is both scientifically accurate and culturally comprehensible.

The Eco Caravan

Two weeks after my visit to the PCC training camp, I had the opportunity to watch the eco caravan perform at two primary schools in two different rural districts of Puerto Princesa City. The audience at each performance consisted of at least fifty children and teachers. In most of the plays, the actors represented local animal and plant species discussing issues of environmental and social relevance, such as conflicts between local fishermen and environmental authorities. Human characters were portrayed as both destroyers (either intentionally or unintentionally) and protectors of nature. The anthropogenic impacts on natural processes were hardly ever questioned. As discussed above, causal connections were often established between natural disasters and environmental behavior that was considered to be morally wrong. This section presents three stage plays that were primarily about climate change.[8] All of them were written and directed by (former) PCC staff members.

---

8   I want to thank Cherry de Dios for providing the scripts of the plays, and Jessa Garibay-Yayen for their translation from Tagalog into English.

"Climate"

The name of the first play presented here is "Climate". It has four main characters—the Sun, the Ozone, Joel, and Vincent—and starts with a dialogue between Sun and Ozone:

> Ozone: Sun, until when will you continue to give off heat?
>
> Sun: I don't know, Ozone. All I know is that this is the role that was given to me by Bathala.[9]
>
> Ozone: Ha, poor humans! They are continually experiencing the constant change in the weather.
>
> Sun: It's their fault, Ozone! They have neglected their duties so much.
>
> Ozone: And what duties are those, Sun?
>
> Sun: What are you thinking, Ozone? You still don't know what they did to you. That is why we have these fluctuations in the weather, like the warming and sometimes the never-ending rain—because of what they did to you!

Ozone does not seem to understand what the Sun is talking about and asks her to explain it. So, the Sun invites Ozone to follow her in order to show how people are responsible for all those weather variations:

> Sun: Come, accompany me to peek into the plains of the world, and we will look for the kind of person which causes the so-called climate change, or frequent weather fluctuations.[10]

On their lookout through the sky they eventually discover two men on an island who illegally cut mangroves with chainsaws. This activity, which is illegal on Palawan, is a source of income for many poor people, who use the wood for charcoal production. Ozone gets very excited, turns to the audience, and says:

> Ozone: Kids, this means that we will find out one of the reasons for the warming of the earth. We should all listen, so that we will understand, okay?

---

9   Bathala is a deity and creator of the universe in pre-Hispanic Filipino mythology.

10 The English term climate change is used here as a literal translation of the Tagalog terms for frequent change (pagbabago-bago) and weather (panahon). Panahon, however, also means season or time.

Both Sun and Ozone approach the men and eavesdrop on them. They become witnesses to how Vincent is rushing his friend Joel, arguing that he doesn't want to get caught by the police and pay a fine. Since it is an unusually hot day, work is getting done slowly. Joel tells Vincent that he is concerned about global warming. He explains to his ignorant friend that cutting mangroves is one of its causes. He notes that it causes droughts to occur, and coral reefs to die. Due to the destruction of the forests there will not be left much to clean "the smoke of the society". The ozone that is supposed to control the temperature of the earth already has holes due to the human use of chlorofluorocarbons.

> Joel: That is why the heat of the sun passes through the ozone. It is no longer controlled by the ozone. Not like before, when it still didn't have holes. It provides the right warmth and coldness that every living creature needs.

Vincent starts to understand. He agrees with Joel that they should change their destructive behavior and find a better livelihood, "like the right way of fishing and vegetable growing, and farming". Joel adds that it would also be a good idea to take part in the community tree planting activities. In the closing scene, Ozone also understands its role in the world now. It turns to the audience and requests to always protect the environment and support its community in environmental activities.

What is very paradigmatic in "Climate" is how closely intertwined the cultural model of climate change and ozone depletion are, which confirms many other studies mentioned above. The common ground for both models seems to be their anthropogenic causes, which are cognitively categorized as environmentally destructive behavior, causing interference with the atmosphere and, therefore, the weather. According to this localized heuristic, the only logical conclusion is to act environmentally friendly in order to stabilize unusual weather patterns.

Fig. 3.3 Eco caravan performing "Climate" in a primary school (photo by author, November 2013), CC BY-4.0.

"The Vegetable Patch of Mr Gorio"

The second play, "Ang gulayan ni mang gorio", is also about climate change and its local impacts. But instead of weather variations, it tells a story about sea level rise, and interestingly this play does not use the English term *climate change* at all. However, its Tagalog translation (*nagbabagong klima*) is used once and the English term *global warming* is used multiple times. The main character, Mr Gorio, owns a small vegetable patch near the sea. He takes care of it as his father and grandfather did before him. One morning, when he leaves his house, he is shocked to find his patch flooded by seawater. He becomes very agitated and calls his wife:

> Gorio: Iska! Iska! Come out here in a hurry! What do you think happened to our garden? Why did the water come in? Was it raining last night, Iska?

Gorio's wife Iska is shocked and starts to cry. All the vegetables seem to be ruined and thus unsaleable in the local market. The family's traditional and, so far, sound livelihood was suddenly at stake. Together they wonder how this could have happened:

> Gorio: How did this happen? Well, it seems this is the first time I've witnessed this. When father was still alive [...], I think it was 1960 (scratches his head), this didn't happen! Why only now? It seems so sudden.

While the family is lamenting about their loss, the village's teacher, Dalisay, comes along. The family immediately beckon her to ask for her professional advice. Dalisay has an answer to what happened to the garden:

> Dalisay: That is an effect of global warming or the heating of the earth.
>
> Gorio: What do you mean, glubal warning?
>
> Dalisay: Not glubal warning (says it with emphasis)! Global warming! The earth is continuously heating up, the ice in the colder places, like the North and South Pole, is melting bit by bit. [...] Ice in the colder places continues to melt, according to scientists. That's why, if they totally melt away, the sea level may be raised higher and now engulfs your garden.

Iska seems to be confused, asking what her vegetables have to do with the melting of ice in distant places. Dalisay explains to her how the oceans are all connected and thus a rising of the sea level is experienced everywhere. In this moment, a new character enters the scene. Buboy, an employee of a non-governmental organization that informs and educates about climate change, tells Gorio and his wife what to do now. He recommends the two IPCC strategies for reducing and managing the risks of climate change—adaptation and mitigation:

> Buboy: Adaptation is about how we can place ourselves in this situation. Do you see that higher part of your land over there? You can move and continue your garden there. That part has been studied and it's been found to be suitable for planting. [...] [Mitigation is] about how we can help reduce global warming. That higher part of your land can still be planted with trees to help here.

On the question of how planting trees will help to fight global warming, Buboy explains that forests are "carbon sinks" that prevent "the smoke" from reaching the atmosphere, where it causes the warming of our planet. Gorio and his wife appreciate the information and agree that they should move their home:

> Gorio: The trees have big roles indeed. I thought they just cradle water and give fresh air, but they also suck smoke. Oh, thanks to you, Buboy and Dalisay.

> Iska: Let it be, and in the next days we shall be moving to that higher ground to live there and continue our garden.

In addition to highlighting once again the important role of trees in climate change adaptation and mitigation, this stage play underpins a basic assumption that I encountered in all my data collections (especially in the pilesorts and interviews), namely that the terms *global warming* and *climate change* were not used interchangeably. Global warming was more strongly associated with global phenomena, i.e., abstract and distant causes like the melting of the polar ice and sea level rise as its consequence. The notion of climate change, however, was much more local and thus directly perceivable.

"Rainy Season in Summer"

Finally, the third play about climate change, "Tag-ulan sa tag-araw" addresses the topic of apparently shifting seasons mentioned earlier. It rarely uses the terms *climate change*, *global warming*, or *sea level rise*, and starts with a fisherman's family that wonders about the strong rainfall that now occurs during the usually dry summer months. The story of the play includes three generations. Whereas Lolo (grandpa) Baste as a person of trust contributes his memories of past times, four children represent a rather skeptical position. The main characters are the children Palaw and Lawan (a play on words referring to Palawan), Luntian (which means the color green in Tagalog), and the skeptical Mina. Further, there are Lawan's parents and his grandfather, who sits in front of his house every morning, looking out to the open sea. One day the kids ask him what he's doing. He replies:

> Lolo Baste: I am waiting for the sun to rise. Because only in this time can I feel the right amount of warmth that it gives us. And look at the bay! Before, the seawater didn't reach our vegetable plantation. Now, it is slowly coming in.

Luntian: How can you say so, Lolo Baste?

Lolo Baste: Because when I was still young, I never witnessed that. But now, it is very different. My dear children—look! Because of the seawater, our crops are dying.

The skeptical Mina counters that the world has always changed and maybe this is just a normal thing to happen. Just as he, Lolo, is not the same man he was decades ago, the sea might change as well. Lolo Baste agrees but adds that this kind of change is different. Even the seasons were no longer what they used to be. In the rainy season (Tagalog: *tag-ulan*), he says, they now have periods of droughts, and the dry season (Tagalog: *tag-araw*) is occasionally affected by heavy rain. Palaw and Lawan note that they have heard about climate change in terms of changing weather patterns in school and on the radio. Mina, however, doesn't buy it. She thinks that maybe today it is a little warmer than in the past but that's all. Lolo Baste explains to her how he used to think like her, but eventually changed his mind. In the past, he admits, he cut mangroves without hesitation, because he was not aware of their important role regarding climate change.

Lolo Baste: That was my belief before. But we were wrong, because the mangroves have a big role in the current change of the climate and global warming. And if we do not bring back the mangroves, I assure you, my dear children, that this problem will be even more serious.

Shortly afterwards, Lawan's father Gusting enters the scene. He came back from his fishing trip much earlier than expected. He looks very sad, so the kids ask him what has happened. Visibly disappointed, he tells them that today's catch was a very bad one. Lawan asks him whether he has an explanation. He answers that according to his experience the movement of the sea has changed. Lawan and Luntian remember that they have also heard something about that:

Lawan: According to the radio, the fishes that we usually see in shallow water are moving towards deeper water. Because what used to be the right amount of heat their bodies can tolerate has changed, and this is because of the warming of the earth.

Luntian: That is not only the effect of the warming of the earth. Because even their homes, coral reefs, are slowly broken into smaller pieces or degraded because the climate is too hot.

Mina remains skeptical. In the neighboring village, she argues, there is still enough fish to buy. The father explains that those fishermen have larger and motorized boats, and that is why they can catch fish in much deeper waters. Compared to them, fishermen with small boats like him lose out.

Suddenly the sky darkens, and bad weather arrives. While the adults run into the house, the children take shelter next to it. After a short discussion, they conclude that Lolo Baste is right, including Mina. Together they decide to do something about climate change. Right after the rain has stopped, they want to collect mangrove seedlings that the storm has washed away and plant them. They hope that the weather will soon stabilize, and fish come back into the shallow waters of their bay again. In a final remark and with special reference to a sustainable lifestyle, Mina turns to the audience and states:

> Mina: That's why we should change! We should love nature and not hurt it! And I hope that it will only be used properly, so the coming generations can still enjoy it, like us. Right, hopefully?

All three eco caravan plays ended with a positive outlook, offering at least one specific action to fight climate change—namely the very popular and socially widespread practice of planting trees, especially in Puerto Princesa City. Institutions such as the eco caravan contribute to maintaining the popularity of tree planting as a means of environmental and climate protection. By distributing knowledge about global warming and climate change to an audience with limited access to such information, the eco caravan can be considered a relevant local translation regime in Puerto Princesa City. The plays paradigmatically represent how the people of Palawan cognitively structure their knowledge about climate change, how they communicate and thus make sense of it. The way climate change is expressed here as a localized, contextualized concept also perfectly mirrors the results of all my other data collections that have been mentioned above (see also Chapter 2, Greenland). Consensually experienced as unusual weather variations, climate change is perceived as a threat to the vulnerable ecosystem of the "last frontier" Palawan that—up to now—has been spared from catastrophic natural hazards, including very strong tropical typhoons. Protecting its lush forest environment is believed to be a vital measure to keep it that way. Although some of the widely shared heuristics are scientifically inaccurate, the benefits of, for example, the thousands of mangroves that have been collectively planted along the coast of the island can hardly be overstated.

Conclusion: Minding the Gap?

This chapter illuminated how climate change is made sense of on the island of Palawan in the Philippines, and the important role played by local processes in translating and reproducing the global discourse on climate change. Following a constructionist approach such as the *sociocultural model* (rather than the *deficit model*) and social representation theory, it was emphasized that sense-making is a multi-layered process that includes cognition and culture. Based on empirical data that was gathered in multi-method fieldwork over several months, it has been demonstrated in this chapter how the people of Palawan make sense of climate change and how they integrate the idea into their local ontologies. By means of various empirical data, it has been shown how climate change is locally perceived and understood: as just another natural disaster that regularly hits the Philippines but hardly ever Palawan, because people there act more environmentally benign. It shows how this sense-making process is inextricably linked to the specific ecological, socio-political, and cultural context of the island; a process that makes climate change a coherent, plausible, and tangible concept that fits into what people already believe, experience and do. The local reception of climate-change discourse therefore depends on pre-existing, shared systems of knowledge and meaning that are reproduced and maintained by circular rather than unidirectional communication. Institutions like the PCSD or the local government administration of Puerto Princesa City are integral components of a complex network of dominant and interwoven discourses and narratives such as the national discourse on natural disasters and Palawan's very popular environmentalism. They all play a role in how knowledge and meaning about climate change are distributed, organized, and structured on the island. The same is true for interpersonal or even performative ways of communication, as the practical example of the eco caravan shows. As instances of local translation regimes, these forms, agents, and interactions of communications more accurately reflect the biogeographical, historical, and legal uniqueness of the island, and everyday experiences of its inhabitants than supra-regional media coverage can. Irrespective of scientific accuracy, local translation regimes therefore much better serve the purpose of translating and communicating climate change knowledge *in a meaningful context*. However, the selectivity of knowledge reproduction generated by national, regional, and local translation regimes (discourse model) should of course be critically examined, especially in terms of the unequal access to knowledge and existing power relations.

The conclusion that can be drawn from this is that the travelling idea of climate change has fallen on very fertile ground on Palawan, where it was easily able to adapt to the local conditions in order to thrive and prosper. The local understanding of climate change and its causalities is widely shared. As Rudiak-Gould found on the Marshall Islands, climate change on Palawan also appears to be an idea that "insults their [island, as it threatens the livelihoods of the inhabitants], but flatters their categories" (Rudiak-Gould 2013a: 177). Basic beliefs and assumptions about human-environment and human-weather relationships are affirmed rather than challenged by the global discourse. It strengthens people's traditionally strong environmentalism and validates their strong rejection of, for example, cutting trees or burning garbage. In accordance with social representations theory (Smith and Joffe 2013), the way people make scientific climate change knowledge familiar and thus comprehensible strongly resonates with pre-existing local terms and cognitive categories. To explore the general question of how climate change is communicated and made sense of in terms of how climate-exposed people understand the scientific concept, the case study of Palawan demonstrates the importance of taking local ecologies and local systems of reference into consideration, i.e., the complex interconnections of climate change knowledge with associated knowledge domains.

In this regard, Palawan is indeed a best practice example of climate change adaptation and mitigation (Department of Environment and Natural Resources 2012). However, this is not the result of a successful communication of scientific climate change knowledge. Rather, the island's success represents an inversion of what is typically meant by the knowledge-behavior or value-action gap in regard to climate change.

Instead of not acting accordingly, the people of Palawan are very committed to preserving and even enhancing their natural environment, and they do it for various reasons. Collectively planting thousands of trees each year is perhaps their most remarkable social practice. While tree planting is framed as a measure to fight climate change, it also fulfils other social functions such as insular identity formation and self-empowerment. In many respects, the people of Palawan do act exceptionally climate friendly. Yet the gap between their apparently climate-friendly behavior and what they know about scientific climate change remains. This shows two implications. First, the *deficit model* and its assumption of a unidirectional relationship between knowledge and action cannot be maintained. Social behavior is complex, and not only driven by explicit knowledge. Second, it shows that a perfectly accurate public understanding of science is not needed to motivate people to implement climate change adaption and mitigation measures. Not surprisingly, even without scientifically tenable beliefs about the environment, weather, and climate, sustainable thinking and behavior are (and have always been) possible.

On Palawan, the discourse on climate change may have served more as post hoc justification than original motivation for past and present behavior, but current anthropological literature on climate change perception suggests that it may not be an exception in this regard (cf. Rudiak-Gould 2013a; Greschke and Tischler 2015; De Wit, Pascht, and Haug 2018). Irrespective of whether people embrace or reject the idea of anthropogenic climate change, the global discourse apparently has the potential to maintain and reinforce local cultures, i.e., pre-existing beliefs, values, and behavioral patterns. From this point of view, it becomes clear why some societies appear to do the right things for the wrong reasons, whereas others appear to do the wrong things despite knowing better. Thus, perhaps the gap between knowing and acting should not be of great concern to climate change communicators who seek to get people involved in climate protection. Rather, they are well advised to adapt to the place and people they address, and to take local human-environment relationships and connected cultural models of nature and the weather into consideration. Instead of focusing only on knowledge transfer, they should keep in mind that local knowledge is always a complex co-product, and that behavioral motivation is derived from various sources, of which scientific knowledge is not known to be the most powerful in the short or medium term. The climate is a slowly changing system, and to a certain extent this also applies to societies. Whilst there is agreement that the constant provision of relevant information based on unequivocal empirical findings is necessary, especially on urgent matters like climate change, there should be more critical assessment and better understanding of how information is translated beyond the realm of science and national policies, and how meanings change between the global and the local levels. In order to transform communication on climate change into a dialogue rather than a one-sided, top-down approach, it is advisable from an anthropological point of view to put more effort into understanding "them" better, before trying to make "them" understand.

References

Baillie, Ian C., Perfecto M. Evangelista, and Nora B. Inciong. 2000. "Differentiation of Upland Soils on the Palawan Ophiolitic Complex, Philippines", *CATENA*, 39.4: 283–99, https://doi.org/10.1016/s0341-8162(00)00078-3

Beck, Ulrich. 2007. *Weltrisikogesellschaft: Auf der Suche nach der verlorenen Sicherheit*, 1st edn (Frankfurt: Suhrkamp).

Bernard, H. R. 2011. *Research Methods in Anthropology: Qualitative and Quantitative Approaches*, 5th edn (Lanham: AltaMira Press)

Blake, James. 1999. "Overcoming the 'value-action gap' in Environmental Policy: Tensions between National Policy and Local Experience", *Local Environment*, 4.3: 257–78, https://doi.org/10.1080/13549839908725599

Borgatti, Stephen P., and Daniel S. Halgin. 2013. "Elicitation Techniques for Cultural Domain Analysis", in *Specialized Ethnographic Methods: A Mixed Methods Approach*, ed. by Jean J. Schensul and Margaret D. LeCompte (Lanham: AltaMira Press), pp. 80–116

Bostrom, Ann, M. Granger Morgan, Baruch Fischhoff, and Daniel Read. 1994. "What Do People Know About Global Climate Change?", *Risk Analysis*, 14.6: 959–70, https://doi.org/10.1111/j.1539-6924.1994.tb00065.x

Bostrom, Ann, and Daniel Lashof. 2007. "Weather or Climate Change?", in *Creating a Climate for Change: Communicating Climate Change and Facilitating Social Change*, ed. by Susanne C. Moser and Lisa Dilling (Cambridge, UK: Cambridge University Press), pp. 31–43, https://doi.org/10.1017/cbo9780511535871.004

Chen, Xiang. 2011. "Why Do People Misunderstand Climate Change? Heuristics, Mental Models and Ontological Assumptions", *Climatic Change*, 108.1–2: 31–46, https://doi.org/10.1007/s10584-010-0013-5

Climate Change Commission. 2011. *National Climate Change Action Plan 2011–2028: NCCAP* (Manila: Climate Change Commission), http://extwprlegs1.fao.org/docs/pdf/phi152934.pdf

Cook, John, Dana Nuccitelli, Sarah A. Green, Mark Richardson, and Bärbel Winkler. 2013. "Quantifying the Consensus on Anthropogenic Global Warming in the Scientific Literature", *Environmental Research Letters*, 8.2: 024024, https://doi.org/10.1088/1748-9326/8/2/024024

Czarniawska, Barbara, and Guje Sevón (eds). 2005. *Global Ideas: How Ideas, Objects and Practices Travel in the Global Economy* (Malmö: Liber & Copenhagen Business School Press)

Czarniawska-Joerges, Barbara, and Guje Sevón. 1996. *Translating Organizational Change*, De Gruyter Studies in Organization 56 (Berlin, New York: Walter de Gruyter).

De Wit, Sara. 2015. *Global Warning: An Ethnography of the Encounter of Global and Local Climate-change discourses in the Bamenda Grassfields, Cameroon* (Leiden: African Studies Centre), https://doi.org/10.2307/j.ctvh9vvwx

De Wit, Sara, Arno Pascht, and Michaela Haug. 2018. "Translating Climate Change: Anthropology and the Travelling Idea of Climate Change", *Sociologus*, 68.1: 1–20, https://doi.org/10.3790/soc.68.1.1

Department of Environment and Natural Resources. 2012. *Climate Change Adaptation: Best Practices in the Philippines* (Manila: Department of Environmental and Natural Resource).

Douglas, Mary (ed.). 1992. *Risk and Blame: Essays in Cultural Theory* (London, New York: Routledge)

Douglas, Mary, and Aaron Wildavsky. 1982. *Risk and Culture: An Essay on the Selection of Technical and Environmental Dangers* (Berkeley: University of California Press)

Eckstein, David, Vera Künzel, and Laura Schäfer. 2017. *Risk Index 2018. Who Suffers Most From Extreme Weather Events?: Weather-Related Loss Events in 2016 and 1997 to 2016* (Bonn: Germanwatch), https://germanwatch.org/de/14638

Engels, Anita, Otto Hüther, Mike Schäfer, and Hermann Held. 2013. "Public Climate-Change Skepticism, Energy Preferences and Political Participation", *Global Environmental Change*, 23.5, 1018–27, https://doi.org/10.1016/j.gloenvcha.2013.05.008

Esselstyn, Jacob A., Peter Widmann, and Lawrence R. Heaney. 2004. "The Mammals of Palawan Island, Philippines", *Proceedings of the Biological Soicety of Washington*, 117.3: 271–302.

Food and Agriculture Organization of the United Nations. 2010. *Global Forest Ressources Assesment: Main Report* (Rome: FOA), http://www.fao.org/3/i1757e/i1757e00.htm

Forest Management Bureau. 2015. *Philippine Forests: Facts and Figures*, 2nd edn (Quezon City: Forest Management Bureau).

Friedrich, Thomas. 2017. *Die Lokalisierung des Klimawandels auf den Philippinen: Rezeption, Reproduktion und Kommunikation des Klimawandeldiskurses auf Palawan* (Wiesbaden: Springer VS), https://doi.org/10.1007/978-3-658-18232-8

——. 2018. "The Local Epistemology of Climate Change: How the Scientific Discourse on Global Climate Change is Received on the Island of Palawan, the Philippines", *Sociologus*, 68.1: 63–84, https://doi.org/10.3790/soc.68.1.63

Glaser, Barney G., and Anselm L. Strauss. 1967. *The Discovery of Grounded Theory: Strategies for Qualitive Research* (Chicago: Aldine Publishing Co.).

Greschke, Heike, and Julia Tischler (eds). 2015. *Grounding Global Climate Change: Contributions from the Social and Cultural Sciences* (Dordrecht: Springer), https://doi.org/10.1007/978-94-017-9322-3

Hastrup, Kirsten. 2015. "Comparing Climate Worlds: Theorising across Ethnographic Fields", in *Grounding Global Climate Change: Contributions from the Social and Cultural Sciences*, ed. by Heike Greschke and Julia Tischler (Dordrecht: Springer), pp. 139–54, https://doi.org/10.1007/978-94-017-9322-3_8

Hulme, Mike. 2009. *Why We Disagree about Climate Change: Understanding Controversy, Inaction and Opportunity* (Cambridge, UK: Cambridge University Press)

Ingold, Tim. 2010. "Footprints through the Weather-World: Walking, Breathing, Knowing", *Journal of the Royal Anthropological Institute*, 16.1: 121–39, https://doi.org/10.1111/j.1467-9655.2010.01613.x

Intergovernmental Panel on Climate Change. 2007. *Climate Change 2007: Impacts, Adaptation and Vulnerability* (Cambridge, UK: Cambridge University Press), https://www.ipcc.ch/report/ar4/wg2/

Jasanoff, Sheila. 2004a. "Heaven and Earth: The Politics of Environmental Images", in *Earthly Politics: Local and Global in Environmental Governance*, ed. by Sheila Jasanoff and Marybeth L. Martello (Cambridge, MA: MIT Press), pp. 31–52.

——. 2004b. "The Idiom of Co-Production", in *States of Knowledge: The Co-Production of Science and Social Order*, ed. by Sheila Jasanoff (London, New York: Routledge), pp. 1–12, https://doi.org/10.4324/9780203413845

——. 2007. *Designs on Nature: Science and Democracy in Europe and the United States* (Princeton, Oxford: Princeton University Press)

——. 2010. "A New Climate for Society", *Theory, Culture & Society*, 27.2–3: 233–53, https://doi.org/10.1177/0263276409361497

Kearney, Anne R. 1994. "Understanding Global Change: A Cognitive Perspective on Communicating Through Stories", *Climatic Change*, 27: 419–41, https://doi.org/10.1007/bf01096270

Kempton, Willett. 1991. "Lay Perspectives on Global Climate Change", *Global Environmental Change*, 1.3: 183–208, https://doi.org/10.1016/0959-3780(91)90042-r

Kempton, Willett, James S. Boster, and Jennifer A. Hartley. 1995. *Environmental Values in American Culture* (Cambridge, MA: MIT Press).

Kreft, Sönke, David Eckstein, Lisa Junghans, Candice Kerestan, and Ursula Hagen. 2014. *Global Climate Risk Index 2015: Who Suffers Most From Extreme Weather Events? Weather-Related Loss Events in 2013 and 1994 to 2013* (Bonn: Germanwatch), https://germanwatch.org/de/11366

Lagmay, Alfredo, Luisa G. Tejada, Rolando E. Pena, Mario A. Aurelio, Brian Davy, et al. 2009. "New Definition of Philippine Plate Boundaries and Implications to the Philippine Mobile Belt", *Journal of the Geological Society of the Philippines*, 64.1: 17–30.

Löfstedt, Ragnar E. 1991. "Climate Change Perceptions and Energy-Use Decisions in Northern Sweden", *Global Environmental Change*, 1.4: 321–24, https://doi.org/10.1016/0959-3780(91)90058-2

Long, Jordan B., and Chandra Giri. 2011. "Mapping the Philippines' Mangrove Forests Using Landsat Imagery", *Sensors*, 11.12: 2972–81, https://doi.org/10.3390/s110302972

Lutes, Mark W. 1998. "Global Climatic Change", in *Political Ecology: Global and local*, ed. by Roger Keil, et al. (London, New York: Routledge), pp. 157–75.

Milton, Kay. 1996. *Environmentalism and Cultural Theory: Exploring the Role of Anthropology in Environmental Discourse* (London: Routledge).

Palawan Council for Sustainable Development. 2004. *State of the Environment: Palawan Philippines* (Puerto Princesa City: Palawan Council for Sustainable Development).

Peterson, Nicole, and Kenneth Broad. 2009. "Climate and Weather Discourse in Anthropology: From Determinism to Uncertain Futures", in *Anthropology and Climate Change: From Encounters to Actions*, ed. by Susan A. Crate and Mark Nuttall (Walnut Creek, CA: Left Coast Press), pp. 70–86.

President of the Philippines. 1991. *Administrative Order No. 220: Creating an Inter-Agency Committee on Climate Change: A.O. 220.*

——. 2007. *Administrative Order No. 171: Creating the Presidential Task Force on Climate Change: A.O. 171.*

Rayner, Steve. 2003. "Domesticating Nature: Commentary on the Anthropological Study of Weather and Climate Discourse", in *Weather, Climate, Culture*, ed. by Sarah Strauss and Benjamin S. Orlove (Oxford, New York: Berg), pp. 277–90.

Republic of the Philippines. 1992. *Republic Act No. 7611–Strategic Environmental Plan for Palawan Act: SEP.*

——. 2009. *Republic Act No 9729–Climate Change Act: CCA*, https://www.lawphil.net/statutes/repacts/ra2009/ra_9729_2009.html

Roncoli, Carla, Keith Ingram, Christine Jost, and Paul Kirshen. 2003. "Meteorological Meanings: Interpretations of Seasonal Rainfall Forecasts in Burkina Faso", in *Weather, Climate, Culture*, ed. by Sarah Strauss and Benjamin S. Orlove (Oxford, New York: Berg), pp. 181–200, https://doi.org/10.5040/9781474215947.ch-010

Roncoli, Carla, Todd Crane, and Ben Orlove. 2009. "Fielding Climate Change in Cultural Anthropology", in *Anthropology and Climate Change: From Encounters to Actions*, ed. by Susan A. Crate and Mark Nuttall (Walnut Creek, CA: Left Coast Press), pp. 87–115.

Rudiak-Gould, Peter. 2012. "Promiscuous Corroboration and Climate Change Translation: A Case Study from the Marshall Islands", *Global Environmental Change*, 22.1: 46–54, https://doi.org/10.1016/j.gloenvcha.2011.09.011

——. 2013a. *Climate Change and Tradition in a Small Island State: The Rising Tide* (New York, London: Routledge), https://doi.org/10.4324/9780203427422

——. 2013b. "'We Have Seen It with Our Own Eyes': Why We Disagree about Climate Change Visibility", *Weather, Climate, and Society*, 5.2: 120–32, https://doi.org/10.1175/wcas-d-12-00034.1

——. 2014a. "Climate Change and Accusation", *Current Anthropology*, 55.4: 365–86, https://doi.org/10.1086/676969

——. 2014b. 'Progress, Decline, and the Public Uptake of Climate Science', *Public Understanding of Science*, 23.2: 142–56, https://doi.org/10.1177/0963662512444682

Smith, Nicholas, and Helene Joffe. 2013. "How the Public Engages with Global Warming: A Social Representations Approach", *Public Understanding of Science*, 22.1: 16–32, https://doi.org/10.1177/0963662512440913

Social Weather Stations. 2013. "First Quarter 2013 Social Weather Survey: 85% of Filipino Adults Personally Experienced the Impacts of Climate Change", June 25, https://www.sws.org.ph/swsmain/artcldisppage/?artcsyscode=ART-20151217092401

Stern, Paul C. 1992. "Psychological Dimensions of Global Environmental Change", *Annual Review of Psychology*, 43: 269–302, https://doi.org/10.1146/annurev.ps.43.020192.001413

Strauss, Sarah. 2007. "An Ill Wind: The Foehn in Leukerbad and beyond", *The Journal of the Royal Anthropological Institute*, 13: 165–81, https://doi.org/10.1111/j.1467-9655.2007.00406.x

Thompson, Michael, and Steve Rayner. 1998. "Risk and Governance Part 1: The Discourse of Climate Change", *Government and Opposition*, 33.2: 139–66, https://doi.org/10.1111/j.1477-7053.1998.tb00787.x

TNS Opinion & Social. 2017. "Special Eurobarometer 459–Wave EB87.1: Climate Change", http://ec.europa.eu/commfrontoffice/publicopinion

UNEP. 2004. *The Global 500 Roll of Honour for Environmental Achievement: 1987–2003.*

United Nations Educational, Scientific and Cultural Organization. 2011. "Philippines: Palawan", http://www.unesco.org/new/en/natural-sciences/environment/ecological-sciences/biosphere-reserves/asia-and-the-pacific/philippines/palawan/

Weingart, Peter, Anita Engels, and Petra Pansegrau. 2000. "Risks of Communication: Discourses on Climate Change in Science, Politics, and the Mass Media', *Public Understanding of Science*, 9.3: 261–83, https://doi.org/10.1088/0963-6625/9/3/304

West, Colin T., and Marcela Vásquez-León. 2003. "Testing Farmers' Perceptions of Climate Variability: A Case Study from the Sulphur Springs Valley, Arizona", in *Weather, Climate, Culture*, ed. by Sarah Strauss and Benjamin S. Orlove (Oxford, New York: Berg), pp. 233–50, https://doi.org/10.5040/9781474215947.ch-013

WWF Philippines, and BPI Foundation. 2014. *Business Risk Assessment and the Management of Climate Change Impact: 16 Philippine Cities*, https://wwf.org.ph/wp-content/uploads/2017/11/BPI-WWF-16-Cities.pdf

Wynne, Brian. 1995. "Public Understanding of Science", in *Handbook of Science and Technology Studies*, ed. by Sheila Jasanoff et al. (Thousand Oaks, London, New Delhi: Sage Publications, Inc), pp. 361–88, https://doi.org/10.4135/9781412990127.n17

4. Sense-Making of COP 21 among Rural and City Residents

The Role of Space in Media Reception

Imke Hoppe, Fenja De Silva-Schmidt,
Michael Brüggemann, and Dorothee Arlt[1]

This chapter explores the role of space in making sense of climate change. The study compares how the United Nations' summit resulting in the Paris Agreement in 2015 was received in an urban (Hamburg) and a rural setting (Otterndorf), both located in Northern Germany. In each setting, two focus group interviews were held (n = 15), one with long-term inhabitants and one with newly relocated citizens. Media coverage was criticized as depicting climate change as overly complex and distant. Use of the local newspaper was more frequent in the rural setting, but its reporting was seen as failing to provide a local angle to the climate summit. Space plays an important role: people in the rural setting—with the rising tides of the North Sea behind the dikes—felt more personally concerned by climate change. Furthermore, long-term inhabitants drew much stronger links between climate change and their region. The duration of stay in a certain setting thus turns out to moderate the influence of space on interpretations of climate change.

1   The research project "Down2Earth" was funded by the German Research Foundation's "Integrated Climate System Analysis and Prediction" (CliSAP) Cluster of Excellence, Universität Hamburg. Further information on the project can be found at URL: https://climatematters.blogs.uni-hamburg.de/down-to-earth/. We would like to thank Aaron McKinnon for his assistance in preparing this manuscript, and Josephine B. Schmitt, Irene Neverla and Katharina Kleinen-von Königslöw for supporting the study and the publication.

https://doi.org/10.11647/OBP.0212.04

Introduction

According to Hulme (2015), there are *many* climate changes around the world: it has different meanings for people in different contexts. How can the emergence of these different meanings of climate change be explained? From a human geography perspective, "space" and "place" are important reference points for explaining these different perceptions and interpretations (Amunden 2015; Ratter and Gee 2012). However, since the "spatial turn" in human geography (Warf and Arias 2009), there has been a consensus that physical characteristics and political boundaries do not dictate certain meanings (Glasze and Mattissek 2012; Ratter and Gee 2012). People living in a vulnerable area are not necessarily more aware of climate change (Lee et al. 2015).

In human geography, communication is understood as a central process in constructing "place". People negotiate the meanings and boundaries of a "place" (Glasze and Mattissek 2012; Agnew 2005). Yet, this is rarely reflected in communication studies (stated as desideratum by, e.g., Autischer and Maier Rabler 2017; Brüggemann et al. 2017; Rodríguez Amat and Brantner 2016; Couldry and McCarthc 2004; Kleinsteuber and Rossmann 1994; Maier Rabler 1992).

Media play a central role in bringing climate change to people's homes—via TV, newspaper, radio, or social networks. This role should be explored in order to better understand climate-related attitudes (Newman, Nisbet, and Nisbet 2018; Taddicken 2013; Brulle, Carmichael, and Jenkins 2012; Arlt, Hoppe, and Wolling 2011).

Content analyses have demonstrated that United Nations (UN) climate summits (known as Conference of the Parties/COP) are a main driver of media attention to climate change (Schäfer, Ivanova, and Schmidt, 2014). These COPs take place as annual meetings of the 194 UN member states that signed the UNFCCC (United Nations Framework Convention on Climate Change). They serve as a focal point and gathering spot for international policy makers and journalists, but little is known about how audiences relate to these events.

The qualitative study presented in this chapter explores how COP coverage is experienced in two different places in order to better understand the extent to which audiences draw a link between media coverage and their living environment. We compare two regions within the same country, so that spatial factors are not overshadowed by substantial differences in culture, language, or socio-economic conditions.

We focus our research on (a) how the *flow of mediated information* about a COP varies in different places and (b) how *media reception and interpretation* of coverage of COPs differs.

The following section will analyze the state of research on how climate change coverage is received, and to what extent spatial aspects play a role. Then, we proceed to describe the research questions and design of the empirical study before discussing the results.

State of Research

A huge body of literature explores how climate change coverage is used, perceived, interpreted, and negotiated (Nisbet 2018). Empirical studies have been conducted in different countries—with a clear bias towards European and US audiences—but the relevance of spatial factors for mediated communication about climate change has not been systematically addressed by communication scholars. Thus, it remains an open question whether (and which) spatial factors influence the process of media reception, and how they may affect climate-change awareness and related constructs.

Past studies have focused on three aspects that will structure our discussion of the literature: (a) (attributed) spatial distance and/or proximity to climate change, (b) personal experiences with nature and weather, and (c) social spaces and group identities.

The Role of (Attributed) Spatial Distance of Climate Change

Does the media's use of proximity (or distance) in depictions of climate change influence the audience's understanding of the issue? Human geography scholars stress that the act of describing an area as a "place of climate change" is a product of social construction (Agnew 2005). Similarly, communication scholars agree that media reporting that classifies a place as being impacted by climate change is a product of mediated social construction (see Neverla et al. 2019). Nonetheless, places "objectively differ in terms of their environmental, social and economic characteristics and these will open up or close down the possibilities open to individuals and groups to interpret proposed place changes" (Manzo and Devine-Wright 2014: 56). The following three studies explore how people perceive and interpret places of climate change through representations in the news media and the effect of (attributed) spatial distance from these places.

O'Neill and Nicholson-Cole (2009) examine fear appeals in media coverage, also with respect to spatial aspects. In their research, focus group discussions in the United Kingdom showed that pictures from a person's direct living environment (like energy-saving lamps or public transport) have positive effects on their intention to act in a climate-friendly manner. Although climate-change awareness and attention to the issue were augmented by 'dramatic visions or human or animal suffering at both local and global scales' (O'Neill and Nicholson-Cole: 371–72), aversive imagery (e.g. starving children, famine) was found to hinder personal engagement because it provokes helpless and overwhelmed sentiments due to the immense dimension of the problem. However, spatial proximity portrayed in pictures that depict negative impacts on "local or regional places that individuals care about and empathize with" (375–76) encouraged people to think about the vulnerability of their living environment to climate change.

Hart and Nisbet (2012) explore whether it matters for a US audience if "climate victims" are depicted as inhabitants of the participants' own country or as living in distant places. The authors use an experimental design to examine the extent to which people's support for climate mitigation policy (dependent variable) increases after reading news articles on the negative health effects of climate change (independent variable). Their mediating variables included spatial distance (climate victims in France versus the US) and group affiliation (Democrats versus Republicans) (Hart and Nisbet 2012: 710). The authors assume that "social identity cues may activate the unintended construct that an issue or problem is not applicable to the group to which a message receiver belongs, and thus the message may be ineffective or result in a negative impact" (705). Their results demonstrate that both mediators indeed influence the media effect—a mechanism the authors call the "boomerang effect". Respondents who strongly identified with a specific US political party interpreted the news article on climate victims in line with their party's official position on climate change: Democrats' support for climate mitigation policy was increased by media exposure of negative climate-change impacts, and Republicans' support decreased even below their baseline values. Hart and Nisbet found that spatial distance also influences perceptions: geographic proximity to climate victims enhances support for climate mitigation policy among individuals who believe climate change is an important problem and decreases support among individuals who deny it. Hart and Nisbet (2012) empirically prove the importance, as well as the complexity, of the role of spatial factors in climate communication. They demonstrate that spatial proximity not only elicits positive responses to mitigation policies; it can also lead to stronger cognitive and emotional defense reactions. This "boomerang effect" did not apply to test subjects who evaluated climate change as an important problem beforehand; for these respondents, geographical proximity reinforced their awareness of the issue.

Jensen (2017) uses "space" as one of four pillars of his theoretical model to analyze how people in Denmark make sense of climate change. A central finding on the space-related dimension is that "world geography enters into both narratives about the origins of climate change and arguments on potential solutions" (449). Audience perceptions of climate change develop in conjunction with definitions of their "space"—e.g. Denmark is perceived as "our little corner of the world" (441), a small country in contrast to a superpower like the United States. In consequence, the Danish participants feel a need to situate themselves in relation to actors in the news of the day, whereas American viewers in a comparable study rarely referred to their own place in world geography because they perceived their nation as the center of the world. Jensen also observed a "North/South" theme in group discussions, related to economic growth and the responsibility of developing countries to also fulfil CO_2 reduction targets.

The Role of Personal Experiences of Weather and Nature

Ryghaug, Sorensen and Naess (2011) use focus group discussions to analyze how Norwegian audiences use weather to contextualize their knowledge about climate change. As one of five "sense-making devices" (784), "nature drama" is a typical interpretive schema that is inspired by and learned through media coverage: "Nature drama events were followed by deliberations with regard to the actual risks. In these exchanges, we noted frequent efforts to soften dramatic media accounts by participants who invoked their own experiences, often to weather" (785). Yet, also the personal experience of "nature drama" is highly interconnected with how media coverage frames the weather.

For Chinese audiences, who have on average a very high climate-change awareness, Wang (2017) found that media usage had no impact on risk perceptions. Instead, a positive correlation between self-reported unusual weather experiences and higher risk perceptions of climate change was found. Brulle, Carmichael, and Jenkins (2012) studied whether extreme weather phenomena in the US (NOAA Climate Extremes Index) shifted public opinion on climate change over a nine-year period (climate change threat index (CCTI)). They concluded that "weather extremes have no effect on aggregate public opinion" (169), but that "the extent of these changes has not reached a level where these shifts can be measured in nationwide polls at the aggregate level. This result may change over time if weather disruptions attributable to climate change increase" (Brulle, Carmichael, and Jenkins 2012: 178; see also Chapter 7, Attribution Science). Indeed, Zanocco et al. (2018) show in a real-life experimental setting that the greater the individual damage experienced in an extreme weather event, and the more this event is discussed in the respective community as attributable to climate change, the more it influences climate-change awareness.

The Social Space:
The Role of Group Identification and Othering

To determine how places are related to interpretations of climate change coverage, perceptions of physical appearances, as well as perceptions and interpretations of the social space (e.g. interpersonal networks like neighborhoods or sports clubs, and hyper-local public spheres like marketplaces or playgrounds) must be considered. Moreover, these social spaces have individual meanings for people, for example in processes both of othering (defining "in-groups" and "out-groups") and identification.

Smith and Joffe (2013: 23) conducted semi-structured interviews with people from the UK, and demonstrate the functions of othering processes: "An 'othering' of the most serious impacts distances the threat by locating it 'out there'". The authors track this view of their UK audiences back to the iconography of mass media coverage, e.g., droughts in Africa, melting glaciers in the Himalaya, polar bears in the Arctic. These processes of othering are thus fostered by media coverage, which is focused on geographically distant places affected by climate change. By attributing climate change to other places of the world, the own "social space" is protected as "a safe place". This outsourcing of the problem is consistent with an assumed vulnerability of others, while being unaffected personally. A similar function of othering is related to the attribution of responsibility: "The antinomy of self versus other plays a core role in everyday conceptualizations of global warming. At one level, the self is regarded as the solution, whereas the other is seen as the perpetrator" (Smith and Joffe 2013: 24).

In these studies, space and proximity are social rather than geographical concepts. Non-physical social spaces influence opinions about climate change and affect the way media coverage is received.

In addition to political affiliation as an instance of group belonging and social space, many other facets could be considered important to understanding the role of social spaces. For example, locality might influence communication flows among meso-structures like sports clubs, schools, choirs, church groups, weekly markets, etc., so these local communities might have an impact on the flow of interpersonal communication, which is crucial to understanding climate communication: "Practices of actively searching for supplementary information, and of engaging in many-to-many communication online about the climate, appear negligible. Instead, one-to-one communication in everyday settings of face-to-face interaction about concrete choices with environmental implications is described as the predominant form of communicative engagement with climate change" (Jensen 2017: 451). How individuals are embedded in geographic and social spaces (e.g. family, sports clubs, colleagues, church groups, political parties, and affiliations like NGOs), and how this affects climate change communication, is a virtually unexplored topic. Othering has been proven to be important, but it remains unclear when (and how) people feel inspired by media coverage to think about their everyday environment as being affected by climate change—i.e. the opposite of othering (in the sense of "Climate change is a topic for us, not only a threat to others"). Table 4.1 summarizes the state of research on space as a dimension to explain the reception of climate change communication.

Table 4.1 Spatial aspects of media reception and effects studies on climate change

Authors	Analyzed Aspects/ Sub-Dimensions	Theoretical Background	Method
(Attributed) spatial distance and/or proximity to climate change, experienced via media			
Hart and Nisbet (2012)	• Spatial distance/proximity towards "climate victims"	• Knowledge deficit model • Motivated reasoning	Experimental design
O'Neill and Nicholson-Cole (2009)	• Spatial distance/proximity combined with "fear appeals"	• Internal control mechanisms of fear (e.g. issue denial)	Focus groups, Q-sorting, semi-structured interviews
Jensen (2017)	• Defining and comparing "here" versus "there"	• Reception analysis	Secondary data analysis with survey data, focus groups
Personal nature and weather experiences as mediated frames			

Ryghaug, Sorensen, and Naess (2011)	• Nature and weather as sense-making devices formed by media experiences	• Domestication of knowledge	Focus groups
Wang (2017)	• Self-attributed unusual weather experiences as consequence of media frames	• Risk perception	Online survey
Brulle, Carmichael, and Jenkins (2012)	• Extreme weather phenomena and their coverage	• Information deficit model, agenda setting, elite cues, etc.	Longitudinal study with a time-series analysis
Social space—othering versus identification			
Smith and Joffe (2013)	• "Othering" as mechanism to distance threats and wresponsibilities	• Social representation theory	Semi-structured interviews
Hart and Nisbet (2012)	• Group affiliation (political partisanship)	• Knowledge deficit model • Motivated reasoning	Experimental design
Jensen (2017)	• Interpersonal communication as key to understanding climate communication	• Reception analysis	Secondary data analysis with survey data, focus groups

Desiderata and Research Question

Even though the studies discussed above did not focus explicitly on the role of spatial aspects in mediated communication about climate change, such aspects appear as important cross-cutting or implicit themes in all cited publications. Current research on climate change communication thus demonstrates that spatial factors do have manifold impacts on how people interpret media coverage of the issue, how they make sense of it, and how media coverage affects their attitudes, knowledge, and behavioral intentions. However, these aspects are analyzed in an isolated, fragmented way or the results were primarily side effects of other research interests.

We seek a better understanding of the role of spatial aspects and how they come into play before, during or after the reception of climate change-related coverage. Our research question is: *What role do spatial factors play in the media reception of COP 21?*

Although this question also calls for comparative international approaches, we start by systematically varying spatial factors and compare an urban versus a rural social setting in the same country. Oriented towards a most similar systems design (MSSD) (Przeworski and Teune 1970), we expect to see how this difference shapes spatial aspects, namely (a) (attributed) spatial distance and/or proximity to climate change, (b) personal nature and weather experiences, and (c) social spaces shape media reception.

Thus, we compare a rather small city (Otterndorf, 7,202 inhabitants) in a rural area with a huge city in a metropolitan region (Hamburg, 1.8 million inhabitants), both of which are located in Northern Germany on the river Elbe and vulnerable to climate change (see below). Hamburg is much further down the river. Otterndorf, while also technically being located on the river, instead evokes the feeling of living right by the North Sea, just behind the dike. Coastal protection is also much less impressive than in Hamburg, with its complex constructions protecting the city against high tides and flooding.

We decided to take a deeper look at COPs as transnational media events (Brüggemann et al. 2017), because they raise media attention across national borders and, potentially, enable audiences to feel connected to each other as global citizens. Content analysis has proven that climate summits are main drivers of media attention to climate change worldwide (Schäfer, Ivanova, and Schmidt 2014). UN climate summits are a main point of reference in climate communication research; most studies focus on communicators on site such as journalists, politicians or NGOs (Lück, Wozniak, and Wessler 2016; Roosvall and Tegelberg 2013; Russell 2013), or on the media coverage (Lück, Wozniak, and Wessler 2016; Zamith, Pinto, and Villar 2013). Yet few studies have explored the audience perspective during climate summits (Wonneberger, Meijers and Schuck 2020). Our research (see also Brüggemann et al. 2017) was conducted during the COP 21 in Paris (2015), which is now seen as a key political event due to its role in renewing and extending the Kyoto Protocol by agreeing on the 1.5°C target (Hulme 2016). Therefore, COP 21 and its corresponding media coverage are not representative of climate summits in general.

Due to the role of COPs as drivers of worldwide media attention, we can make sure that during COP 21 the issue was covered by the media in both regions analyzed. However, the salience of the issue in the individual media repertoires of participants in both places is an open question. The first sub-question thus addresses the patterns of information and news usage in both places:

- **Sub-Question (a). Patterns of Information:** How do people's media use and interpersonal communication about COP 21 differ in both locations?

As patterns of information and communication inform the interpretation and appropriation processes, they are important to consider as they provide contextual information for the interpretation of sub-question (b), which is our main research interest:

- **Sub-Question (b). Patterns of Interpretation:** How do people interpret media coverage of climate change with regard to spatial distance, the natural environment they live in and their social space?

Because local discourses are nourished by physical, interpersonal presence, they foster face-to-face communication (Mettler-Meibom 1992). As a cross-cutting theme, we focus on the role of direct personal communication as a bridge between media and local discourses on climate change.

Method: Comparing Rural and City Residents

This section compares the two cases with respect to their geographic proximity to projected climate threats and selected conditions that shape the social space and are relevant to media reception and effects (especially the presence of local media). Next, their differences are outlined. The last part of this section describes the method of focus groups and media diaries, and presents the data sources analyzed in this chapter.

Similarities between the Two Study Locations

Hamburg and Otterndorf are both strongly influenced and shaped by the North Sea and the River Elbe (see Fig. 4.1). While Otterndorf is situated directly on the coast at the extreme end of the Elbe estuary, Hamburg lies further upstream on the Elbe. At the position of Hamburg, the tides and tidal range are still very strong. Both regions are therefore perceived as being potentially vulnerable to climate change effects, especially a higher frequency and potentially more severe impacts of storm surges due to the expected sea level rise (Meinke et al. 2010). The whole region has experienced grave floods in the past such as the North Sea flood of 1962, which has become part of the collective memory in the region (Trümper and Neverla 2013; NLWKN 2017a). The region has also been recently affected by severe storm surges. In November 2007, windstorm Tilo hit the German bight (St. Pauli/Hamburg: NN + 5,42 m; Cuxhaven/Otterndorf: NN + 4,44 m); in December 2013 it was Xaver (St. Pauli/Hamburg: NN + 6,09 m Bremerhaven/Otterndorf: NN + 4,99m).

Fig. 4.1 Locations of Hamburg and Otterndorf. © OpenStreetMap contributors, CC BY-SA.

Storm Xaver in 2013 produced "one of the most severe storm surges of the past hundred years", raising the water level by three meters in some areas (NLWKN 2017b: author's translation). The dike system, which is built along the entire North Sea coast, prevented injuries and major property damage in both study areas.

Fig. 4.2 Impact of storm surges. Above: Otterndorf at the beginning of
the storm surge in 2013 (photo by Ekkehard Drath, CC-BY-NC). Below:
Hamburg during a storm surge in 2016 (photo by Olle August, from
Pixabay, https://pixabay.com/photos/
high-water-hamburg-port-motifs-3930235/).

Otterndorf is directly exposed to the North Sea, and as part of the
Wadden Sea Region (WSR) it is especially at risk from sea level rise
(Gerkensmeier, Ratter, and Vollmer 2016). Coastal researchers have
called for the development of an integrative risk management approach
for the region.

Although Hamburg is less directly exposed to the North Sea, it would still be at risk during storm surges *without* its flood protection system; seventeen of its districts are officially categorized as risk areas. In the city's most dire risk scenario—an extreme storm surge combined with a collapse of the protection system—33,000 persons would be directly affected (Freie und Hansestadt Hamburg 2013).

In Germany, climate-change awareness of the public is quite high. Accordingly, we found in the quantitative survey data that both regions have a similarly high climate-change awareness, which does not differ from the average population in Germany (Brüggemann et al. 2017). While 55% of the country's population agrees that "global climate change is a very serious problem", fewer than 20% feel "very concerned it will personally harm them during their lifetime" (Stokes, Wike, and Carle 2015). A yearly telephone survey administered to approximately 500 Hamburg inhabitants from 2008 to 2018 has generated quantitative data about residents' risk perceptions and personal concerns about climate change (Ratter 2018). In contrast to the objective risk assessments discussed above, 40% of respondents do not believe they will be affected by any natural catastrophes such as storm flooding, heatwaves or heavy rainfall. The results even show a decline in risk perceptions for Hamburg from 2008 to 2011. However, no clear trend can be observed. Unfortunately, these data are not available for rural residents along the German North coast.

Ratter and Gee (2012) describe *Heimat* as a German concept of regional perception and identity. For the Wadden Sea Region, "Natural resources play a strong role in shaping people's perception of Heimat. Specific mentions included resources such as the Wadden Sea, beaches, the coastal landscape, nature on the coast, fresh air, as well as fish and seafood, all of which are used by the people living on the coast in many different ways" (131). As both locations are part of the Wadden Sea Region, one can assume that this description also applies to Hamburg and Otterndorf, although Hamburg is less exposed to the sea than Otterndorf.

In both locations, local print media play an important role. In Hamburg, the *Hamburger Abendblatt* has a large distribution; also, many hyperlocal (online) newspapers are published. In Otterndorf, the *Niederelbe Zeitung* has a lower circulation — due to the lower number of residents — it is complemented by the weekly newspaper *Hadler Kurier*. Table 4.2 compares the two locations.

Table 4.2 Spatial aspects indicating similarities between
Otterndorf and Hamburg

Indicator	Hamburg and Otterndorf	Source
Exposure to climate change	• Strongly influenced and shaped by the North Sea and the River Elbe • Potentially vulnerable to climate change effects, especially a higher frequency and potentially more severe impact of storm surges due to the expected sea level rise	(Meinke et al. 2010)
Coastal protection	• Dike systems	(Müller and Gönnert 2016)
Shared features of *Heimat* understandings	• Natural resources like water and fresh air are important, as well as food (fish) and the coastal landscape	(Ratter and Gee 2012)
Local media	• *Hamburger Abendblatt* (167,000 copies daily) • *Niederelbe Zeitung* (7,400 copies daily) • *Hadler Kurier* (25,800 copies per weekly issue)	(Informationsgemeinschaft zur Feststellung der Verbreitung von Werbeträgern e.V. 2018)
Public climate-change awareness ("climate of opinion")	• Both regions are located in Germany—where climate awareness is quite high on average but people feel personally affected by climate change only to a lesser degree	(Brüggemann et al. 2017; Stokes, Wike, and Carle 2015)

Differences between the Two Locations

While Hamburg is Germany's second-largest city with a high population density and approximately 1,78 million residents (Statistisches Amt für Hamburg und Schleswig-Holstein 2019), Otterndorf is a small town in a rural area with 7,202 inhabitants (Landesamt für Statistik Niedersachsen 2017). Hamburg has a huge but less dense social network that makes it possible to stay isolated. Otterndorf is located in a rural area with a low population density (Stadt Otterndorf 2016).

In the literature it is assumed that in rural settings, "everyone knows everyone", and that social connections are closer than in the city: "[C]ompared to personal networks in urban settings, personal networks in rural settings contain ties of greater intensity and role multiplexity, are based more on kinship and neighborhood solidarities rather than on friendship, are smaller, are denser, and have greater educational, race-ethnic, and religious homogeneity, but less age and gender homogeneity" (Beggs, Haines, and Hurlbert 1996: 306).

Cities in general have a far more heterogeneous population than smaller communities, which can lead to the formation of sub-cultures and sub-networks in which traditional ties (e.g. relatives, neighbors) no longer play a role: "community, notably urbanism, also affects the social contexts from which people draw their relations, albeit far more weakly than social class, life cycle, and other personal traits" (Fischer 1982: 79-80).

Coastal protection plays an important role in climate change adaptation. Hamburg has a EUR 600 million flood protection system consisting of mechanical barriers and a central warning system (called KATWARN). Detailed risk maps define the dangers and protection measures for the neighborhoods close to the river (Müller and Gönnert 2016). The system has been constructed with future upgrades in mind, required due to sea level rise (Freie und Hansestadt Hamburg 2012). Otterndorf, in contrast, has not recently renewed its protections, the most important of which is the dike. Some of its pumping stations and floodgates have been in use for decades (Niedersächsischer Landesbetrieb für Wasserwirtschaft, Küsten- und Naturschutz) or even centuries (the Otterndorf floodgate was put into service in 1854 and last reworked in 1985, Kosch 2020).

Table 4.3: Spatial aspects indicating differences between
Otterndorf and Hamburg

Indicator	Hamburg	Otterndorf	Source
Social structure	• Approx. 1,78 million inhabitants • Mean age: 42.75 years	• Approx. 7,202 inhabitants • Mean age: 21–64 years (60 %)	(Statistisches Amt für Hamburg und Schleswig-Holstein 2019; Stadt Otterndorf 2016)
Assumptions regarding social networks	• High population density, "big city life"—huge, but less dense social network	• Low population density, rural area where "everyone knows everyone"—close(r) connections, dense networks	(Beggs, Haines, and Hurlbert 1996; Fischer 1982)
Status of coastal protection	• Mechanical flood protection system with capacities for further extensions	• Central mechanical barriers in need of renewal	(Müller and Gönnert 2016)
Financial situation of coastal protection measures	• High financial capacities for new protective measures	• No recent investments	(Freie und Hansestadt Hamburg 2012)

Method

This study is part of a larger research project at the Universität Hamburg called "Down to Earth"[2] (Brüggemann et al. 2017). It analyzes how the media coverage of COP 21 reached and affected citizens in Germany. The project combines three interconnected sub-studies with different methodological approaches: a quota-based online panel survey (n = 1,121) including three waves, digital communication diaries (n = 42), and four focus group interviews (n = 15). All sub-studies have a

---

2   The project was funded by the German Research Foundation's "Integrated Climate System Analysis and Prediction" (CliSAP) Cluster of Excellence, Universität Hamburg. Further information can be found at https://climatematters.blogs.uni-hamburg.de/down-to-earth/

common interest in shedding light on media reception of climate politics. The qualitative parts of the project allow us a deeper understanding of how people make sense of COP 21, while the standardized survey allows to draw broader conclusions about public perceptions of climate policy in Germany. In this chapter, we concentrate on material from the focus groups and media diaries and contextualize the results with data obtained from the online survey.

Focus group discussions: We conducted four focus group discussions—two in Otterndorf (n = 9) and two in Hamburg (n = 6)—during the final week of COP 21. The conference was on the media agenda, but the focus groups took place before the final press conference, so the outcome of the conference did not overshadow participants' memories of the entire coverage. Participants were aged 30 to 75. A "contrasting case selection" sampling strategy (Flick 2007) was used, which aims to vary the central aspect of theoretical interest so that alternative explanations can be found, and thereby helps to explore the research question as openly as possible. If spatial aspects play a central role in the reception of climate change communication, the results from both groups should vary, at least in parts.

On the basis of the literature analysis we assume that it is not only the physical space that effects how people interpret media coverage on climate change, but also the social aspects of space, for example a feeling of "*Heimat*" [belonging], the degree of connectedness to others, the set of experiences made at a certain place (e.g. floodings), etc. Thus, for the focus groups we varied the degree of connectedness people have to their living spaces. We were not able to realize pre-interviews to fully understand these aspects a priori. Instead, we focused on a more objective criterion, which is the length someone had lived at her/his current address. We grouped participants in each location in two homogenous groups, one with "long-term inhabitants" and one with "short-term inhabitants" (living less than six years in the current place). Additionally, a rough quota target for diversity regarding age (min: 30; max: 75), gender (5 w: 10 m) and professions (e.g., administration employee, electro technician, translator, pensioner, teacher) was implemented.

The collected data was analyzed via a qualitative content analysis (Schreier 2013), and categories were built inductively and deductively.

The initial classification mainly contains the two dimensions detailed in sub-research questions (a) patterns of information and communication and (b) patterns of interpretation.

Communication diaries: During the two weeks of COP 21, 42 participants from Hamburg and Otterndorf filled out digital communication diaries on a daily basis. In these online diaries, they noted their daily media use and face-to-face conversations about climate change. To encourage steady participation, the respondents received a financial incentive. The focus group participants were also part of the diary survey.

Three-wave online panel survey (n = 1.121): A three-wave online panel survey was administered throughout the event: the baseline measurement was conducted two weeks before COP 21, the main survey was carried out during the conference, and a final survey four weeks after.

Results

This section interprets the findings of the focus groups and media diaries; the results of the online panel study are described to contextualize the statements of the participants. Quotes are translated, and participants' names are pseudonyms.

Patterns of Information

During the conference, media diary participants reported each day whether they had noticed media coverage of the COP 21 or on climate change in general. Compared to the baseline study that covered the German population (results published in Brüggemann et al. 2017), the findings for the participants of the study in Hamburg and Otterndorf demonstrate that people received information about the climate conference through their usual media repertoires, and that the amount of information did not differ by location (see Table 4.4); participants from both places recalled an average of ten media contacts about COP 21. However, some of their regular sources did not seem to contribute news on the topic at all. For example, while private television news is popular for rural participants, they hardly noticed climate change reporting there (on 0.3 days on average during COP 21).

Table 4.4 News reception in Hamburg and Otterndorf during COP 21

Medium	Hamburg M (SD)	Otterndorf M (SD)
"Do you remember one or more news items on climate change today?" (summative index from 0 to 13 days)		
News and information on public television (incl. online)	2.5 (2.7)	2.7 (1.8)
News and information on private television (incl. online)	0.1 (0.3)	0.3 (0.8)
News and information on radio (incl. on-line)	2.0 (2.6)	1.4 (2.0)
Printed national newspaper	0.4 (1.2)	0.1 (0.3)
Printed weekly magazine	0.2 (0.5)	0.0 (0.0)
Printed local newspaper*	**1.0 (1.6)**	**2.5 (2.1)**
Bild-Zeitung (tabloid press)	0.0 (0.0)	0.0 (0.0)
Online newspaper	2.7 (2.8)	2.5 (2.4)
Online portal (like google news)	0.6 (1.1)	0.1 (0.3)
Social networks	0.2 (0.5)	0.4 (1.4)
Blogs and online discussions	0.3 (1.0)	0.0 (0.0)
Video platforms	0.1 (0.3)	0.0 (0.0)
Total number of news contacts (on average)	10.10	10.00

Note: n = 13 for Otterndorf; n = 29 for Hamburg; summative index: added the answer (yes = 1; no = 0) for the entire enquiry period (30 November—12 December 2015). Significant differences between Hamburg and Otterndorf are printed in bold, *p ≤ .05.

Respondents in both places received information from public television and online newspapers most often (on average on almost three out of thirteen days, see Table 4.4), followed by radio (De Silva-Schmidt and Brüggemann 2019). Individual variances between respondents were high in both locations. Some respondents only mentioned two or three different sources of media information during the entire enquiry period (mostly a print newspaper and public radio or TV). Others named a variety of online sources such as TEDTalks on Youtube, science blogs, and media websites, while using traditional media at the same time.

This wide variance in our sample also explains the high standard deviation for most items and might be a reason why there is only one significant difference: the use of local print newspapers. About two-thirds of respondents (both younger and older) read local papers at least sometimes. While Hamburg residents read information on climate politics in their local news on only one day during the entire study period, Otterndorf participants noted an average of two-and-a-half days. Interestingly, none of the participants who recently moved to Otterndorf or Hamburg read a local newspaper, suggesting that exposure to local news is connected to how long a person has lived in their current area. As we show below, this is also likely to lead to a less localized perception of climate change.

During the climate summit, residents from Hamburg and Otterndorf also mentioned a lot of interpersonal communication in their media diaries.

Table 4.5 Interpersonal communication during COP 21

Medium	Hamburg M (SD)	Otterndorf M (SD)
"Do you remember talking about climate change today?"		
... with family and friends	1.2 (1.5)	2.2 (2.5)
... with colleagues and acquaintances	0.4 (0.6)	0.7 (1.1)

Note: n = 13 for Otterndorf; n = 29 for Hamburg; summative index: added the answer (yes = 1; no = 0) for the entire enquiry period (30 November—12 December 2015). Index scales ranges from min = 0 days to max = 13 days. There are no significant differences between groups.

In both places, there were many "non-talkers"—sixteen in total, which again explains the high standard deviation and is possibly a reason why there was no significance in the t-test. However, of those who did talk about climate change during the COP 21, regional differences are notable.

First, the *number of conversations* about climate change differs. Respondents from Otterndorf (n = 13) noted 47 individual conversations (3.6 per person) about climate change during the enquiry period, which is over one-and-a-half times more conversations per person than the participants from Hamburg (n = 29, talks: 59). Otterndorf residents not only talked more often about climate change, but they also had more diverse interlocutors, conversing with family and friends as well as acquaintances and colleagues about the topic

Second, the *topic of the conversations* varied. Participants in Hamburg talked more about the climate conference, about technological measures to prevent climate change (*"carbon capture and storage"*, Niklas, twenty-eight years old), and about the endangering effects on nature in general, such as the extinction of vulnerable species. The group discussions revealed that Hamburg participants discussed climate change more often in the context of their professions (e.g. in the energy sector) or lifestyle choices (e.g. with their flat-sharing community). In contrast, respondents in Otterndorf talked more often about the negative effects of climate change on their living environment (i.e. high tides, floods and heavy rains, also having other consequences such as climate refugees) and local adaptation measures. Many of them discussed their personal responsibility. Helge, a sixty-two-year-old long-term resident of Otterndorf, declared: "We need to start with our own consumption behavior. If everyone does it, we are many".

For the Otterndorf participants, interpersonal communication also served the purpose of sharing and trading knowledge, especially on local issues such as dike protection and storm surges. Max, 31 years old, recently moved to Otterndorf, said: "I heard our colleague [...] speak. And he said that [...] behind the dike not everything is as safe as we all believe. I don't know what area he referred to". Florian, thirty-three years old, recently moved to Otterndorf, replied: "Altenbruch? [a neighboring village]". Max: "Yes, he said it's just a matter of time before it [the dike] collapses".

With reference to the section on differences between the two regions, Max's assumption fits with the objective data about Otterndorf's vulnerability and the quality of the dikes, some of which are obsolete and in need of renewal. This example demonstrates that in Otterndorf, interpersonal communication serves as an important medium to ask for, spread and negotiate information on local issues such as dike quality and climate change adaption. It thus compensates for a perceived lack of reporting from local journalism. The participants asserted that this is an urgent issue not covered in local newspapers or radio stations. Manuel, forty-one, who has lived in Otterndorf for six years, reported: "I have been asking myself in the last few days how the quality of the dike actually is, and the opinions differ very much on how good the dike here in Otterndorf is at all. [...] There are also voices that claim that the dikes here are not that good".

Elderly Otterndorf participants assumed that dike protection and storm surges were not of interest to younger people. According to Anton, a sixty-four-year-old long-term resident:

> Yes, I think for the older generation in particular it is even more [a topic] than for the younger generation. Because [the older generation] saw that from childhood on, what the storm surges were like, or what it means to have floods. Or if the dikes start breaking, we'll have seen it all here. [...] One has enormous fears, yes. Young people haven't even experienced it: "Oh, the dike is safe". They don't care at all.

Indeed, in the second group with younger participants, Otterndorf respondents assessed that climate change was irrelevant. Florian, thirty-three years old, recently moved to Otterndorf, assumed: "Yes, maybe we will have a higher sea level. But we have great dikes, so to me this discussion is not relevant".

Hamburg participants also mentioned that climate change in general is not salient enough for private conversations. Participants from both locations reported that climate change is not an issue in their circle of friends and colleagues. Dirk, sixty-two years old, reported: "I notice that in my personal discussion, with friends or colleagues, it [the refugee crisis, high numbers of refugees coming to Germany due to the war in Syria in 2015] plays a greater role than climate".

The participants also assume that climate change does not bring enough news value for private conversations. Dirk explains: "In my circle of friends, a real debate about the topic is not possible because the problem is universally known", and Stefanie (thirty-eight, Hamburg) added "We have more serious problems at the moment here, for Hamburg the application for the Olympic Games". Anton (sixty-four, Otterndorf) stated similarly: "...because you just don't feel the immediate threat, right? That is, I think, how many people think. It's just hard to keep thinking about future generations".

Patterns of Interpretation

The first part of this section discusses the common patterns of interpretation in both locations. When asked about their perception of news on COP 21, the following patterns are characteristic of the reception and interpretation: a perceived neglect of the issue in media coverage, complexity, abstractness, and single key moments of being overwhelmed and having obtained new insights.

Weak Salience of the Issue in Broadcasting Media—Disappointment with Local News

From a scholarly perspective, it is often assumed that climate change is a tough topic for journalism because it does not interest broad audiences (see, e.g., Jensen 2017). Yet, we find that the respondents complained about *too little* media coverage of climate change on TV and radio. Participants from both locations criticized journalistic coverage for providing too short and superficial news items—a complaint that is widely confirmed by the country-wide survey (n = 1,121) (De Silva Schmidt and Brüggemann 2019; Brüggemann et al. 2017). For example, 84% stated that the media did not report enough on the COP 21 negotiations, 85% agreed that the media did not report appropriately on political parties' conflicts and strategies, and 82% accused the media of reporting insufficiently on climate change in general. Margret (sixty-two, Otterndorf) complained about "insufficient coverage in public service broadcasting", and Sandra (forty-five, Hamburg) noted "far too few pieces of information". The opposite was true for national daily newspapers. Fewer participants read them, but those who did reported that they provide much-appreciated detailed insights (e.g., Margret: "The article was really captivating and well-written"), although others mentioned not having enough time to read them properly: "Unfortunately, I only had time to skim through" (Luisa, twenty-six, Hamburg).

The local print media, however, were rated extremely poorly, especially by the long-term residents—e.g. by Anton, sixty-four, long-term resident in Otterndorf: "I think it is so important, that the local newspapers in particular, that they regularly supply their readers with what happens there [at the COP 21 conference]—but this just wasn't the case!"). Both groups assessed their local newspaper as biased and negative, using the sparse coverage of COP 21 as an example. Participants worried about other people in their city or region who use the local newspaper as their single source of information, supposedly elderly people ("I worry about people who do not have the opportunity to read online news", Anton, Otterndorf). Economic interests of the publisher were held responsible for the poor performance of the local Hamburg newspaper. Elisabeth, sixty-two, and a long-term resident of Hamburg, criticized: "In our obedient and submissive local newspaper there was nothing on COP 21. In comparison to e.g. *Süddeutsche Zeitung*, our local newspaper in Hamburg is a provincial newspaper". Here, the spatial reference ("provincial") was used to negatively connote the failure of local newspapers to convey the local angle of climate change. They do not connect what happened at COP 21 in Paris to Hamburg, but more importantly they do not connect the elite discourse on climate change to local discourses, and thereby fail to create a social space of communication.

In summary, our first result is that participants would have liked to receive more information on climate change and COP 21. This finding is confirmed by the survey data and is thus a general pattern of reception of COP 21 in Germany (Brüggemann et al. 2017), and not limited to the cases of Otterndorf and Hamburg. Especially regarding their local newspapers, many respondents were disappointed about the low salience of the issue in their own local media outlet and the perceived failure to connect the issue to local discourses.

Complexity and Abstractness— Climate Change and the COP as Distant Issues

One reason for the complaints about too little information on COP 21 in media coverage could be rooted in the perceived high level of complexity of the issue. Local audiences were unable to make sense of the multiple facets of climate change and of climate politics. Respondents referred to the complex terminology used in media coverage, and describe it as hard to understand and to translate into their everyday language (Anton: "Very, very complex issue, I had to learn a lot in order to be able to pursue the discussion. And the terminology, 2 °C target and the like").

Moreover, they criticized cumbersome and long-winded printed news articles, which are not understandable and readable within a normal amount of time. "Abstractness" is a related concept, but it comes with the feeling of not being able to grasp the meaning of climate change for one's own daily life.

Participants described the COP 21 as far away from their everyday life reality (Anton, sixty-four, Otterndorf,: "These climate summits are relatively far away. We need a concrete example on the ground to make sense of climate change. So, for us, this would be the dike"). This statement demonstrates that the perceived distance from the event in Paris has a cognitive, emotional and social dimension. Instead of receiving "abstract" information, participants would have liked to learn about the consequences climate change has for their local environment. In addition, Hannah (forty-three, Hamburg) reported: "In my circle of friends, I feel there is a big distance from the topic". Here, the spatial metaphor is also employed to describe the basic feelings and cognitions about climate change. Distance may refer less to geographic distance and more to the perceived disconnect between climate change and one's personal life within its social and physical space.

Being Overwhelmed and Affected by Media Coverage— Moments of Connection

Although the participants described a fundamental sense of distance, touching moments of very personal connection were reported as well— when the media coverage opens up a social space of genuine encounter. A sense of "being overwhelmed" is very salient in our focus group data. Also, participants of the group discussions felt touched by the coverage on rare occasions—not as a persistent feeling, but in extraordinary moments.

Triggers can be media persona (fictional and non-fictional characters) who are perceived to be socially close and directly affected by climate change in the near future, e.g. Barbara (sixty-two, Otterndorf) said "What touches me the most is if people tell how they are affected by climate change, e.g. a farmer who said that he will be able to farm another two years but after that his land will be washed away". These can also be people who act as a personification of powerful values, here in the form of Native Americans: "When Indians appear in their outfits and demand (a change) from industrialized countries" (Barbara), who are perceived as ambassadors bridging various forms of cognitive and emotional distance.

The "happy ending" of the COP in Paris was much reported in the media diaries and perceived as a touching moment: "Everyone waited all day long for the result and talked about how it would be. Then when the news of the signing came, that was great" (Margret, sixty-two, Otterndorf). Others felt "joy and hope after the conference" (Birgit, fifty-three, Hamburg) as well as relief: "Which I did not dare to hope... after twenty-three years finally a working approach, unfortunately still (with) many compromises and perhaps also too late" (Susanne, forty-nine, Hamburg). Here, another reason for the described distance from the issue came to the surface: due to many failed previous conferences, and an inability at the policy level to reach progress, the topic was considered too frustrating. Distance thus also serves as a mechanism of "self-protection" from bad news, such that the "good news" encouraged people to re-connect.

The dimension "getting new insights" includes a positive experience as well: learning new facts during COP 21—even ones that question ones' own beliefs—was evaluated as positive. Dirk (sixty-two, Hamburg) reported:

> I have already learned a lot during the COP, especially about the interests of individual states. So, for example [...] this smog situation in Beijing, then you (thought before) "Oh God, thank God it's not like that here", but what I did not know is that China, for example, does a lot for climate protection, which does not yet have the addressed effects, but they are very active there.

Climate reporting was positively assessed if it facilitated insights into geographically and socially distant countries to which the respondents otherwise had no access.

"We are Vulnerable" versus "We are a Leading Industrial Nation"

A main finding with respect to the relevance of spatial aspects for reception and interpretation is that long-term residents were more likely to link COP 21 coverage to their region or city.

Moderator: "Did you think about Otterndorf while you watched coverage of the climate summit?" Anton, sixty-four: "Yes, yes, yes. It is a huge topic for us". Klaus, sixty-eight, long-term resident in Otterndorf, immediately added "I have a dinghy in my garden!"

However, this connection is sometimes, again, followed by mechanisms of suppression.

Anton: "You have to care for yourself to not get depressed. Otherwise you cannot live with that fear", and Martin, forty-six, agreed: "There's always a certain fear, that's for sure. And we assume it [the dike] will hold. Otherwise we would not make any new developments in this area, we would not have built us a house".

For long-term Otterndorf residents, fear and serious concerns about their area seemed to be enormous, yet rather hidden, as an everyday life issue. Klaus, sixty-eight, long-term resident of Otterndorf:

Yes, climate change is a fast-moving issue, especially if you consider the connection to infrastructural projects like Elbe dredging [with consequences for the height of storm surges], that is also a huge topic for us here. It's all connected. And that's why we all should walk around here every day complaining [about our situation]. But at some point you get depressed. [Agreement from the entire group.] If you can't do anything and still have the danger on the back of your neck.

The Otterndorfers described themselves as affected by huge infrastructure projects such as dredging to deepen the Elbe, to support Hamburg's huge harbor industry. This economically motivated infrastructure project places an additional strain on the dike system, on top of climate change. The participants reported that the dredging has already made floods and storm tides more destructive.

For the long-term Hamburg residents, the harbor and its responsibility for climate change (e.g., emissions of cruise liners) were the main reference point in responding to the question if Hamburg would be directly affected by climate politics. Elisabeth, sixty-two, stated that "economy and jobs in Hamburg are reasons put forward, especially cruise liners as most dirty polluters like Queen Mary". Local media coverage is evaluated as extremely unsatisfying and disappointing by the long-term residents of Hamburg, due to a perceived lack of coverage of the manifold meanings of climate change for the city. They especially miss a particularly critical investigation: "There was once a report, that is really long ago, and it was about the 'Hafencity' [the newly built harbor district], because of the air pollution the Hafencity performed so poorly!" (Elisabeth, sixty-two). Some new residents of Hamburg assume that the media's lack of localizing COP issues for the city indicates that climate change is not a risk for Hamburg and that there is no reason to worry. Karl (forty, Hamburg): "There is no relevance for Hamburg, otherwise the media would have had reported on that".

Many participants, both from Otterndorf and Hamburg, viewed the aim of COP 21 as engaging "other countries" in climate policy, assuming that Germany already had a pioneering role.

Max, thirty-one, newly moved to Otterndorf: "Germany already has a leading position in climate protection; COP 21 is more about engaging other countries".

They attribute responsibility for climate change to other nations' governments and declare that Germany is a leading industry and technology nation that cannot be affected by climate change. Othering was applied to qualify their country's responsibility against those of others; moreover, Germany is seen as completely untouched by the effects of climate change due to technical fixes. Max asserts:

> Here in Germany, we are lucky. Because we have great dikes, the coverage on COP 21 and its effects on Otterndorf are irrelevant to me. Even if a dike breaks down, we'll build a new one. So, we are lucky in Germany. They can't do that on those small islands in the Pacific. So, there the sea level rises, and then they go down.

Similarly, Karl, forty, a new inhabitant of Hamburg, stated: "I do not see a link between COP 21 and Hamburg. Consequences of climate change happen elsewhere. Even if the River Elbe rises about a metre, so what? We can handle that. In other countries, the situation is much more severe. Droughts, lack of water. One storm more or less, we will survive". This statement shows how he creates a perceived distance between his home environment and "other countries", "elsewhere", which can be seen as a case of othering.

The study participants exhibit the same pattern of thinking as in the panel survey: There has to be more engagement on climate protection, but someone else, somewhere else (other countries, the politicians) should act (Brüggemann et al. 2017). However, it is striking that people who have lived in the same place for a long time—in both locations—think and talk much more about their personal responsibility and that of their community than those who recently moved to their area. "There are a lot of grassroot initiatives in this region, and they organize, for example, local country markets to sell products from local farms". Klaus, sixty-eight: "Yes, but that is not the majority. These initiatives are noticed by a minority only". Margret, sixty-two, Otterndorf added: "Indeed, these initiatives are not our typical everyday life—that is our supermarket".

References to the Physical Environment— Similar but Different

The most prominent climate change concern for all respondents is flooding and dike protection. For Hamburg, the harbor is an equally important point of reference. Residents of Hamburg and Otterndorf differ in the extent to which they are personally concerned about potential climate-change impacts. Respondents from Hamburg seem to feel that climate change is unlikely to concern them personally, or only in an abstract way since it might affect humanity as a whole. One participant (Karl, forty, short-term resident) reported: 'I didn't see a direct connection to Hamburg [...] because the really painful effects will take place somewhere else. So, even if the Elbe rises a meter, that's still all doable". Although Hamburg is potentially vulnerable to flooding, participants from the city did not believe they were likely to be personally affected by storm surges or other climate change effects. Even though "land submerged in the North" (Annette, forty-five, short-term resident) is a tangible consequence of climate change for them—it will not affect them personally. According to sixty-two-year-old Elisabeth, a long-term resident: "That Wilhelmsburg [suburb of Hamburg heavily affected by the 1962 storm surge] is flooded again, that can't happen to us today. So, I had the feeling, yes, we can sit back and relax now, and we are also warned in time [...] to drive away our car".

Hamburg residents perceive the effects of climate change and sea level rise at an emotional distance (which some respondents refer to as "suppression" of the disturbing knowledge); problems only seem to exist elsewhere, e.g. "in Africa" (Birgit, fifty-three, long-term resident). Yet, for respondents from Otterndorf, high tides in the Elbe and floods are dangers that are easy to grasp and represent a direct threat to their physical environment. Thus, a major difference is that Hamburg participants feel much less threatened by climate change than Otterndorf participants, who are located much closer to the North Sea and have several weak points in their dike systems.

Although the participants from Otterndorf also mention serious consequences for faraway regions such as the Fiji Islands, they paint a much clearer picture of how climate change might affect their hometown. They talk about past experiences of "high tides in the Elbe" (Margret, sixty-two, long-term resident), and future "storm flooding warnings for our coastal region", and even think about the possibility of losing their home in the future: "in case of a sea level rise, there might be a wave of refugees coming from the North Sea region" (both quotes by Anton, sixty-four, long-term resident in Otterndorf). They also specifically refer to themselves: "We talked about how climate change is connected to our town, what we can do, what has already changed. We just became aware that this topic also concerns us" (Margret, sixty, long-term resident).

Conclusion

The study compared media reception and sense-making of news on climate change in a big city versus a small rural town located directly on the coast. As Hamburg and Otterndorf are located close to each other in Northern Germany, it is not surprising that common patterns of both media use and interpretations of climate change emerge. Yet, we found that spatial factors play an important role in both (a) how news about climate change and policy making are received, and (b) how people engage and make sense of it.

Sub-Question (a): The findings demonstrate that people received information about the climate conference COP 21 through their usual media repertoires, as the quantitative survey data and media diaries show. However, many regular media sources provided little information about the event and the issue. By testing group differences statistically, it turns out that the amount of information received did not differ by location, so that differences in perception and interpretation of both groups (sub-question b) cannot have been influenced hereby. Many of the regular sources—such as private television news— did not seem to contribute news on the topic at all. Rural residents

(Otterndorf) significantly accessed more local sources of news (i.e., local newspapers), as demonstrated by the survey data and media diaries. They are engaged with the issue more closely by talking more about climate change, and they connect climate change more to their direct living environment. Interpersonal communication plays a more important role in the rural setting: personal conversations serve to pass on knowledge about vulnerable aspects of the area, especially between new and old residents. It is reasonable to assume that living right next to a dike in a rural setting with more first-hand experience of weather phenomena makes the issue of climate change more salient.

Sub-Question (b): We found common patterns related to sense-making of news on COP 21. Spatial references abound among both groups of participants. A perceived weak salience of the issue in mass media coverage is echoed by the perception of a weak salience in personal discussions. The study participants are disappointed by the perceived failure of local media to account for the importance of the negotiations in Paris for their region. Climate change and climate policy are perceived as complex and abstract and the media are blamed for not explaining them well. Thus, local publics experienced the issue of climate change and the COP 21 negotiations as distant events. Yet, media personas and some iconic personal stories, even if being from other parts of the world, are able to bridge the perceived social and geographical distance and establish an emotional connection with climate change and policy making.

   These differences in information pathways go hand in hand with different interpretations of climate change. Residents from the rural setting and from the city experience the spatial distance to climate change slightly differently. Many of the spatial aspects identified in previous studies in the literature could also be found in our data. Although climate change is an abstract and "far-away" concept for all of our respondents, personal concerns were more prominent in the rural setting and connected to a feeling of potentially being personally affected. For participants from Otterndorf, climate change was more relevant to one's personal life and fears. Respondents from Hamburg felt less personally concerned and saw themselves less as potential victims but (partly) as responsible for causing climate change. These different feelings are connected to different nature and weather experiences, and the variation in the strength of dikes and protection systems in both places leads to different risk perceptions of the inhabitants. Some participants domesticated climate change in interpersonal discussions by referring to the current weather or the natural environment.

An important finding is that the differences presented were more salient for the long-term residents in our sample. People who are more rooted in the area they live in seem to perceive climate change as more severe. Local newspapers are mainly used by those who have lived in an area for longer, and connections between climate change and one's own living environment were mostly drawn by long-term residents. It seems that the longer a person lives in a place and the more connected he or she feels to it, the more relevant spatial factors become for his or her experience of climate change. The time spent living in one place, and the extent of being rooted in that space, seems to be an important variable that mediates the influence of space.

This mediating factor emerging from our study has important implications in a world that has become increasingly mobile and less rooted in physical space: the more people move around, the less they will access local media, and the less they will know about (or be interested in) their local area and how climate change might affect it. They might be more inclined to view climate change as a distant issue that has no implications for one's personal life. If the local media, for those long-term residents who use them, are also failing to draw the connection between global warming and local development, this might constitute a barrier to initiate and promote adaptation and mitigation processes on the local level—but this would have to be tested by future research.

Talks about climate change also reflect the economic inequality between the city, that is better shielded against floods and the (perceived) neglect of dikes in the rural area. Climate change may deepen the (perceived) inequality between wealthy and important cities and the far less privileged countryside. Local media fail in the perception of our interviewees to critically discuss these problems. The failure of local media is partly compensated through face-to-face discussions, particularly among long-term inhabitants who seem to be more inclined to draw the link between global policy-making at COP 21 and local politics.

References

Agnew, John. 2005. 'Space: Place', in *Spaces of Geographical Thought: Deconstructing Human Geography's Binaries*, Society and Space Series, ed. by Paul Cloke and Ron Johnston (London: SAGE Publications Ltd), pp. 81–96, https://doi.org/10.4135/9781446216293.n5

Amundsen, Helene. 2015. "Place Attachment as a Driver of Adaptation in Coastal Communities in Northern Norway", *Local Environment*, 20.3: 257–76, https://doi.org/10.1080/13549839.2013.838751

Arlt, Dorothee, Imke Hoppe, and Jens Wolling. 2011. "Climate Change and Media Usage: Effects on Problem Awareness and Behavioural Intentions", *International Communication Gazette*, 73.1-2: 45–63.

Autischer, Alfred, and Ursula Maier-Rabler. 2017. "Kommunikationsatlanten als Konzept für eine raumbezogene Kommunikationsforschung", *MedienJournal*, 9.2: 27–32, https://doi.org/10.24989/medienjournal.v9i2.1251

Beggs, John J., Valerie A. Haines, and Jeanne S. Hurlbert. 1996. "Revisiting the Rural-Urban Contrast: Personal Networks in Nonmetropolitan and Metropolitan Settings", *Rural Sociology*, 61.2: 306–25, https://doi.org/10.1111/j.1549-0831.1996.tb00622.x

Brüggemann, Michael, Fenja De Silva-Schmidt, Imke Hoppe, and Josephine B. Schmitt. 2017. "The Appeasement Effect of a United Nations Climate Summit on the German Public", *Nature Climate Change*, 7.11: 783–87, https://doi.org/10.1038/nclimate3409

Brulle, Robert J., Jason Carmichael, and J. Craig Jenkins. 2012. "Shifting Public Opinion on Climate Change: An Empirical Assessment of Factors Influencing Concern over Climate Change in the U.S., 2002–2010", *Climatic Change*, 114: 169–88, https://doi.org/10.1007/s10584-012-0403-y

Couldry, Nick, and Anna McCarthy. 2004. *MediaSpace: Place, Scale and Culture in a Media Age* (London, New York: Routledge), https://doi.org/10.4324/9780203010228

De Silva-Schmidt, Fenja, and Michael Brüggemann. 2019. 'Klimapolitik in den Medien—das Publikum erwartet mehr.: Befunde einer Befragung zu den UN-Klimagipfeln 2015 und 2018', *Media Perspektiven*, 50.3: 107–13.

Fischer, Claude S. 1982. *To Dwell Among Friends: Personal Networks in Town and City*, 4th edn (Chicago: University of Chicago Press)

Flick, Uwe (ed.). 2007. *Designing Qualitative Research* (London: Sage Publications, Ltd), https://doi.org/10.4135/9781849208826

Freie und Hansestadt Hamburg. 2012. *Sturmflutschutz in Hamburg gestern – heute – morgen*, 10th edn (Hamburg: LSBG) https://www.hamburg.de/pressearchiv-fhh/3286334/2012-02-09-bsu-sturmflutschutz/

———. 2013. *Umsetzung der Hochwasserrisikomanagementrichtlinie in Hamburg: Information der Öffentlichkeit gemäß § 79 Wasserhaushaltsgesetz (WHG) über die Umsetzung der Hochwasserrisikomanagementrichtlinie (Richtlinie 2007/60/EG) in der Flussgebietsgemeinschaft Elbe* (Hamburg) https://www.hamburg.de/contentblob/4146070/dd0201b68660a4025a0feed391020096/data/download-begleittext.pdf

Gerkensmeier, Birgit, Beate Ratter, and Manfred Vollmer. 2016. "Trilateral (Flood) Risk Management in the Wadden Sea Region", in *ENHANCE. Novel Multi-Sector Partnerships in Disaster Risk Management*, ed. by Jeroen Aerts and Jaroslav Mysiak (Brussels: European Commission), pp. 210–31, https://www.technopolis-group.com/wp-content/uploads/2020/04/Enhance-Layout-WEB-low.pdf

Glasze, Georg, and Annika Mattissek. 2012. *Handbuch Diskurs und Raum: Theorien und Methoden für die Humangeographie sowie die sozial- und kulturwissenschaftliche Raumforschung*, 2nd edn (Bielefeld: Transcript).

Hart, Philip S., and Eric C. Nisbet. 2012. "Boomerang Effects in Science Communication: How Motivated Reasoning and Identity Cues Amplify Opinion Polarization About Climate Mitigation Policies", *Communication Research*, 39.6: 701–23, https://doi.org/10.1177/0093650211416646

Hulme, Mike, and Martin Mahony. 2010. "Climate Change: What Do We Know about the IPCC?", *Progress in Physical Geography: Earth and Environment*, 34.5: 705–18, https://doi.org/10.1177/0309133310373719

Hulme, Mike. 2015. "Climate and its Changes: A Cultural Appraisal", *Geo: Geography and Environment*, 2.1: 1–11, https://doi.org/10.1002/geo2.5

———. 2016. "1.5 °C and Climate Research after the Paris Agreement", *Nature Climate Change*, 6: 222-24, https://doi.org/10.1038/nclimate2939

Informationsgemeinschaft zur Feststellung der Verbreitung von Werbeträgern e.V. 2018. "Quartalsauflagen ab 2016", https://www.ivw.de/aw/print/qa

Jensen, Klaus B. 2017. "Speaking of the Weather: Cross-Media Communication and Climate Change", *Convergence*, 23.4: 439–54.

Kleinsteuber, Hans J., and Torsten Rossmann. 1994. *Europa als Kommunikationsraum: Akteure, Strukturen und Konfliktpotentiale in der europäischen Medienpolitik* (Opladen: Westdeutscher Verlag).

Kosch, Andreas. 2020. "Hadelner Kanalschleuse Otterndorf", https://www.nlwkn.niedersachsen.de/hochwasser_kuestenschutz/landeseigene_anlagen/schleusen/schleuse_otterndorf/kanalschleuse-otterndorf-41074.html

Landesamt für Statistik Niedersachsen. 2017. "Bevölkerung und Katasterfläche in Niedersachsen", https://www.statistik.niedersachsen.de/themenbereiche/bevoelkerung/themenbereich-bevoelkerung---tabellen-87673.html

Lee, Tien M., Ezra M. Markowitz, Peter D. Howe, Chia-Ying Ko, and Anthony A. Leiserowitz. 2015. "Predictors of Public Climate-change awareness and Risk Perception around the World", *Nature Climate Change*, 5.11: 1014–20, https://doi.org/10.1038/nclimate2728

Lück, Julia, Antal Wozniak, and Hartmut Wessler. 2016. "Networks of Coproduction: How Journalists and Environmental NGOs Create Common Interpretations of the UN Climate Change Conferences", *The International Journal of Press/Politics*, 21.1: 25–47, https://doi.org/10.1177/1940161215612204

Maier-Rabler, Ursula. 1992. "In Sense of Space: Überlegungen zur Operationalisierung des Raumbegriffs für die Kommunikationswissenschaft", in *Zeit, Raum, Kommunikation*, ed. by W. Hömberg and M. Schmolke (Munich: Ölschläger), pp. 357–70.

Manzo, Lynne C., and Patrick Devine-Wright (eds). 2014. *Place Attachment: Advances in Theory, Methods and Applications* (London: Routledge).

Meinke, I., et al. 2010. "Klimawandel in Norddeutschland: Bisherige Änderungen und mögliche Entwicklungen in Zukunft", in *Kalte Zeiten-Warme Zeiten: Klimawandel(n) in Norddeutschland*, ed. by mamoun Fansa and Carsten Ritzau, pp. 56–59.

Mettler-Meibom, Barbara. 1992. "Raum – Kommunikation – Infrastrukturentwicklung aus kommunikationsökologischer Perspektive", in *Zeit, Raum, Kommunikation*, ed. by W. Hömberg and M. Schmolke (Munich: Ölschläger), pp. 387–401.

Müller, Jan-Moritz, and Gabriele Gönnert. 2016. *Entwicklung der Hochwasserschutzanlagen in Hamburg* (Hamburg: LSBG), https://www.klimacampus-hamburg.de/fileadmin/user_upload/klimacampus-aktiv/160420_Sturmfluttagung/MuellerGoennert_Hochwasserschutzanlagen.pdf

Neverla, Irene, Monika Taddicken, Ines Lörcher, and Imke Hoppe (eds). 2019. *Klimawandel Im Kopf: Studien Zur Wirkung, Aneignung und Online-Kommunikation* (Wiesbaden: Vieweg), https://doi.org/10.1007/978-3-658-22145-4

Newman, Todd P., Erik C. Nisbet, and Matthew C. Nisbet. 2018. "Climate Change, Cultural Cognition, and Media Effects: Worldviews Drive News Selectivity, Biased Processing, and Polarized Attitudes", *Public Understanding of Science*, 27.8: 985–1002, https://doi.org/10.1177/0963662518801170

Niedersächsischer Landesbetrieb für Wasserwirtschaft, Küsten- und Naturschutz. "Unterhaltung von landeseigenen Anlagen—dafür steht der NLWKN", https://www.nlwkn.niedersachsen.de/hochwasser_kuestenschutz/landeseigene_anlagen

Nisbet, Matthew (ed.). 2018. *The Oxford Encyclopedia of Climate Change Communication* (New York: Oxford University Press), https://doi.org/10.1093/acref/9780190498986.001.0001

NLWKN. 2017a. "Auch die höchsten Deiche geben keine absolute Sicherheit: Rückblick auf die Sturmflut 1962", https://www.nlwkn.niedersachsen.de/hochwasser_kuestenschutz/kuestenschutz/rueckblick_auf_sturmfluten/rueckblick_auf_sturmflut_1962/auch-die-hoechsten-deiche-geben-keine-absolute-sicherheit-rueckblick-auf-die-sturmflut-1962-102700.html

———. 2017b. "Nikolausflut vom 6. Dezember 2013", https://www.nlwkn.niedersachsen.de/hochwasser_kuestenschutz/kuestenschutz/rueckblick_auf_sturmfluten/sturmflutbilder_2013/nikolausflut-vom-6-dezember-2013-120645.html

O'Neill, Saffron, and Sophie Nicholson-Cole. 2009. "'Fear won't do it': Promoting positive engagement with climate change through visual and iconic representations", *Science Communication*, 30.3: 355–79, https://doi.org/10.1177/1075547008329201

Przeworski, Adam, and Henry Teune. 1970. *The Logic of Comparative Social Inquiry* (New York: John Wiley & Sons, Inc.).

Ratter, Beate M. W. 2018. "Wahrnehmung des Klimawandels: Veränderung der Wahrnehmung", *Klimanavigato*, https://www.klimanavigator.eu/dossier/artikel/037475/index.php

Ratter, Beate M. W., and Kira Gee. 2012. "Heimat—a German Concept of Regional Perception and Identity as a Basis for Coastal Management in the Wadden Sea", *Ocean and Coastal Management*, 68: 127–37, https://doi.org/10.1016/j.ocecoaman.2012.04.013

Rodríguez-Amat, Joan R., and Cornelia Brantner. 2016. "Space and Place Matters: A Tool for the Analysis of Geolocated and Mapped Protests', *New Media & Society*, 18.6: 1027–46, https://doi.org/10.1177/1461444814552098

Roosvall, Anna, and Matthew Tegelberg. 2013. "Framing Climate Change and Indigenous Peoples: Intermediaries of Urgency, Spirituality and De-Nationalization", *International Communication Gazette*, 75.4: 392–409, https://doi.org/10.1177/1748048513482265

Russell, Adrienne. 2013. "Innovation in Hybrid Spaces: 2011 UN Climate Summit and the Expanding Journalism Landscape", *Journalism*, 14.7: 904–20, https://doi.org/10.1177/1464884913477311

Ryghaug, Marianne, Knut Holtan Sorensen, and Robert Naess. 2011. "Making Sense of Global Warming: Norwegians Appropriating Knowledge of Anthropogenic Climate Change", *Public Understanding of Science*, 20.6: 778–95, https://doi.org/10.1177/0963662510362657

Schäfer, Mike S., Ana Ivanova, and Andreas Schmidt. 2014. "What Drives Media Attention for Climate Change? Explaining Issue Attention in Australian,

German and Indian Print Media from 1996 to 2010", *International Communication Gazette*, 76.2: 152–76, https://doi.org/10.1177/1748048513504169

Schreier, Margrit. 2013. *Qualitative Content Analysis in Practice* (Los Angeles: Sage).

Smith, Nicholas, and Helene Joffe. 2013. "How the Public Engages with Global Warming: A Social Representations Approach", *Public Understanding of Science*, 22.1: 16–32, https://doi.org/10.1177/0963662512440913

Stadt Otterndorf. 2016. *Nordseebad Otterndorf—Die Grüne Stadt am Meer: Wohnen, Leben & Arbeiten* (Otterndorf: Nordseebad Otterndorf), https://www.stadtmarketing-otterndorf.de/fileadmin/user/broschueren/imagebroschuere-otterndorf-leben-wohnen-arbeiten.pdf

Statistisches Amt für Hamburg und Schleswig-Holstein. 2019. "Monatszahlen—Bevölkerung", https://www.statistik-nord.de/zahlen-fakten/bevoelkerung/monatszahlen/

Stokes, Bruce, Richard Wike, and Jill Carle. 2015. "Global Concern about Climate Change, Broad Support for Limiting Emissions: U.S., China Less Worried; Partisan Divides in Key Countries", *Pew Research Center: Global Attitudes & Trends*, 5 November, https://www.pewresearch.org/global/2015/11/05/1-concern-about-climate-change-and-its-consequences/

Taddicken, Monika. 2013. "Climate Change from the User's Perspective", *Journal of Media Psychology: Theories, Methods, and Applications*, 25.1: 39–52, https://doi.org/10.1027/1864-1105/a000080

Trümper, Stefanie, and Irene Neverla. 2013. "Sustainable Memory. How Journalism Keeps the Attention for Past Disasters Alive", *Studies in Communication and Media*, 1: 1–37, https://doi.org/10.5771/2192-4007-2013-1-1

Wang, Xiao. 2017. "Understanding Climate Change Risk Perceptions in China: Media Use, Personal Experience, and Cultural Worldviews", *Science Communication*, 39.3: 291–312, https://doi.org/10.1177/1075547017707320

Warf, Barney, and Santa Arias. 2009. *The Spatial Turn: Interdisciplinary Perspectives*, Routledge Studies in Human Geography 26 (London, New York: Routledge).

Wonneberger, Anke, Marijn H. C. Meijers, and Andreas R. T. Schuck. 2020. "Shifting Public Engagement: How Media Coverage of Climate Change Conferences Affects Climate Change Audience Segments", *Public Understanding of Science*, 29.2: 176–93, https://doi.org/10.1177/0963662519886474

Zamith, Rodrigo, Juliet Pinto, and Maria E. Villar. 2013. "Constructing Climate Change in the Americas: An Analysis of News Coverage in U.S. and South American Newspapers", *Science Communication*, 35.3: 334–57, https://doi.org/10.1177/1075547012457470

Zanocco, Chad, Hilary Boudet, Roberta Nilson, Hannah Satein, and Hannah Whitley. 2018. "Place, Proximity, and Perceived Harm: Extreme Weather

Events and Views about Climate Change", *Climatic Change*, 149.3–4: 349–65, https://doi.org/10.1007/s10584-018-2251-x

5. What Does Climate Change Mean to Us, the Maasai?

How Climate-Change Discourse is Translated in Maasailand, Northern Tanzania[1]

Sara de Wit

This chapter explores the varying ways in which the Maasai pastoralists in Terrat village in northern Tanzania give meaning to climate-change discourses. This study moves away from the idea that there is a "linear" (from global to local/science to citizen) and "correct" way of interpreting and understanding climate change as a scientific discourse, but turns the question around by asking "what does climate change mean to the Maasai"? Based on fourteen months of multi-sited ethnographic fieldwork, this chapter contextualizes climate-change discourses in the historical, environmental and political dimensions of the Maasai's "interpretive horizons". It is argued that local discourses and interpretations are not just barriers in the global pursuit for climate change adaptation, even if they contradict global discourses and policies, but reveal crucial insights about local priorities, values, and agency. In other words, the rejection of this new discourse should not be seen as a form of ignorance, but rather as an act of cultural translation and resistance.

1   This chapter is based on research that I carried out in 2012–13 for my PhD at the University of Cologne, and is a reworked version of Chapter 8 of my dissertation (de Wit 2017). My research was funded by the DFG, part of the SPP1448 priority programme, and supervised by Michael Bollig and Dorothy Hodgson, to whom I am greatly indebted.

https://doi.org/10.11647/OBP.0212.05

Introduction: Climate-change discourse Dawns on Terrat

In northern Tanzania, where the Maasai pastoralists dwell, a nascent discourse about a changing climate is dawning on the rural village of Terrat. This new story resonates with the experience of some herders, who have observed a decline in rainfall levels, which they attribute to a decreased sense of morality in society. Others who have heard about "this thing called climate change" find this new scientific explanation unconvincing, or utterly confusing, since they see climate variability as old news. After all, the semi-arid plains of their homeland have always been prone to climatic extremes and fluctuations. The elderly in particular report that they have not observed unusual changes in the weather or climate recently, and they emphasize that the region has always experienced "bad years".

Villagers who listen to the radio in Terrat hear about a global problem that is already affecting their locality, which will only become more severe in the future. And they are told that the main culprits of the looming crisis are the rich countries with their polluting industries, but that the Maasai—and all the other poor people in the world—will suffer the most. However, they are also told (mainly by their own government) that they are part of the problem, because they cut down trees for firewood and have too many cattle. They are encouraged to plant trees, and explain that carbon dioxide is dangerous, that God has nothing to do with it, that there is something called an ozone layer protecting the earth that humans are depleting and that science is real (more real than God), among many other things.

While the new prophecy is disseminated through multiple sources, such as non-governmental organizations (NGOs), researchers (like myself), the church and the radio, I observed an overall lack of climate change information, awareness and conversation.[2] Furthermore, while

---

2   The main language in Terrat is Maa, the Maasai language, which is spoken by the whole Maasai population, whereas the level of Swahili is generally dependent upon education, connectivity and exposure to external influences. There is one radio station in Terrat, which is the only radio station in Tanzania that broadcasts programmes in both Swahili and Maa. Many villagers own a mobile phone, but few have a smartphone, so internet access is very limited. Despite recent efforts to give women access to radios, they are still predominantly owned and listened to by men.

many informants indeed complained about the lack of rain, people's day-to-day worries did not center on the global climate crisis; they had more pressing concerns like a fear of losing their land, having access to basic health care, vaccinations for their animals, education and secure access to natural resources like land, water and grasses. Environmental hardships, such as recurring droughts and irregular rainfall, exacerbate this complex set of challenges (De Wit 2019). Based on fourteen months of multi-sited research carried out in Tanzania, this chapter discusses how climate-change discourse is translated, communicated and "received" in Terrat village.[3] While much of the anthropological literature has focused on "observing" climate change (how communities perceive climate change directly as a sensory process), and a few studies have explored "receiving" (the uptake of scientific information through secondary sources), scant attention has been paid to interactions between the two (Rudiak-Gould 2013). In line with Rudiak-Gould, in this chapter I argue that understanding how people make sense of climate-change discourse requires taking both observation and reception into account. This is not only because people's environment and climate play an important role in how this new discourse is ascribed meaning; new discourses about a changing climate also shape the ways in which people talk about their environment (De Wit 2019; Rosengren 2018; De Wit 2015; Rudiak-Gould 2011). The first and most obvious reason for the lack of climate-change awareness—as a scientific and anthropogenic discourse—is a lack of access to information and education. However, my research made it clear that this "void" must be understood in light of Maasailand's complex historical, political, environmental, and ontological dimensions that comprise the broader interpretive context through which people make sense of climate-change discourses (De Wit 2017). But there is more to this absence, since there are villagers who have heard about climate change but were not impressed by (or were indifferent to) its apocalyptic undertone. This chapter explores

---

3   While the concept of "receiving" climate change is commonly used in reception studies (as proposed by Rudiak-Gould 2011), it has a somewhat passive connotation (the term "reader response" as used in hermeneutics overcomes this problem). I prefer to use the notion of translation, which denotes agency—a process in which the producer and reader of a text make an equal contribution.

this complex mixture of absence, translation, observation, and reception (including rejection). It investigates how the climate-change discourse is interpreted in Terrat and entangles old and new horizons, and how it is at times embraced by some, but refuted by others.

In dominant models of risk perception, communication and management, found in fields like natural hazard research and psychometric risk studies, a passive recipient of an independent stimulus is implied, as it is essentially concerned with information transmission and how to educate the recipient with minimum distortion (Rayner 1992: 85). While this insight is not new in media and cultural studies (see, e.g., Hall 1973), in relation to climate change research the preoccupation remains largely concerned with the question how to tailor messages to audiences (Nerlich 2017; Moser 2016, 2010). Both cultural theory and reception studies (in the modern hermeneutical tradition), diverge from such conventional theories of risk perception and the perceiver of information, or "the reader", is treated as an active and social agent. Furthermore, against a common reductionist trend that exists in the social sciences and policy literature that views religion (and at times culture) as barriers in the larger project of climate change adaptation (e.g. Haraway 2016; Kuruppu and Liverman 2011; Paton and Fairbairn-Dunlop 2010; Taylor 1999), I explore the meaning-making processes as emergent knowledge spaces in their own right (Hastrup 2015) that emerge when climate change is received by the Maasai in Terrat. I argue against the "purification" (Latour 1993) of scientific and religious knowledges but rather focus attention to the ways in which people navigate different registers as they make sense of climate change and the agentive possibilities it affords. The disregard for both cultural but even more so religious perspectives in the literature emerges from a misunderstanding and secular rejection of religious thought, underpinned by "the desire to enforce the boundaries between the religious and the scientific" (Fair 2018: 5; Kempf 2017; Hulme 2014). In this chapter, I argue that the rejection of this travelling discourse should not be perceived as a form of ignorance, but as resistance—an attempt to remain faithful to one's own set of norms, values, beliefs, principles of causality and "cosmological configuration" (Hermann and Kempf 2018; Fair 2018: 9; Kempf 2017: 34; Rudiak-Gould 2014, 2013). The (partial) refusal and other ways that appear to disrupt the hegemonic framing of climate science can be seen as a form of agency in its own right (Fair 2018: 9; Hermann and Kempf 2018; Kempf 2017: 34).

In this chapter, I examine what happens when a rural village like Terrat is confronted with a new narrative about a changing world that carries a message of doom and decay. How do villagers, who have never seen industries, receive and translate this story for which they hold no responsibility, and within which God is relegated to the margins? The following account provides insights into a village that is confronted with a new (and still somewhat alien) discourse, which brings new explanatory pathways about a changing world and climate into being. I conclude by advancing the argument that translating this new discourse can only be meaningful for the Maasai if consideration is given to the question what climate change means to them.

Climate-change discourse is here referred to as "a set of various concepts, models and representations that comprises of scientific information about climate and climate change, which undergoes continuous translation by an array of translators such as scientists, journalists, governments, NGOs, activists, anthropologists, local communities etc." (De Wit et al. 2018: 3). It is important to point out that climate science is not a monolithic but a vast and heterogeneous body of knowledge that consists of an array of varying perspectives, positions, methods, internal translations, etc. (cf. Dürr and Pascht 2017: 3). Furthermore, climate science (or climate scientists for that matter) is not necessarily secular, but this dichotomy might emerge in the translation of a "modern" discourse vs. a "traditional" one. I also do not assume an opposition between science and other "modes of existence" but scrutinize the emergent meaning making processes (Latour 1993). As philosopher of religion Berry has observed, while not being a novel insight per se, "over the past two decades many anthropologists have contributed to a critique of purportedly 'objective' scientific ways of producing environmental knowledge, arguing that, like the cosmologically grounded vision of many Indigenous societies, so too are the ecological ideas of Euro-American cultures colored by religious elements" (Berry 2016: 78), or underpinned by different "myths" of nature (Rayner and Heyward 2014; Hulme 2009).

Fig. 5.1 Terrat in Simanjiro District (Manyara Region). Cartography by Monika Feinen, produced in November 2015, CC-BY-ND.

Translating Climate Change in Terrat

In an educational short movie clip called "Climate Conscious Programme", made by several NGOs and recorded in Terrat with a group of villagers, Maasai pastoralist Leboi[4] is herding cows in the Simanjiro Plains.[5] He and his fellow "actors" are lamenting about a changing world:

> Actor 1: There are changes in this world and I don't know why.

> Leboi: There has been a prolonged drought, at least three years with little rain. Even those who came here from Kenya had their cattle die because of the drought.

> Actor 2: So where will I give my cows water?

Then Emmanuel, a documentary filmmaker, sits with them under a tree and begins to explain in Maa:

---

4   To ensure confidentiality, the names of informants and organizations may have been changed or left unmentioned.

5   This movie was produced by an international NGO together with Tanzanian based organizations. The movie is scripted, so that is why I refer to these Maasai herders as actors. Translation from Maa was provided by documentary makers.

Emmanuel: Let me tell you something. I learned about this at school. We are not the only ones experiencing these changes. It is like if you pinch yourself on the finger. You will feel the pain all over. It is the same for the world.

Then Emmanuel takes a football that symbolizes the globe. He continues his explanation:

Emmanuel: Let's say this is the world we are living on.

Actor 2: That ball?

In the next shot the ball has turned into a globe.

Emmanuel: There is a place called America, which is here (points it out on the globe). And a place called Europe, here, and another one called China. These places produce a lot of smoke because they have a lot of vehicles, and airplanes, and they have these things called industries [see Fig. 5.2]. They burn a lot of fuels, like diesel, petroleum and oil. After they burn, the smoke comes out, which is called carbon dioxide. That is the bad one. The pollution rises into the sky where it affects our protective blanket. Then the blanket becomes denser and denser. And the whole world becomes hot.

Leboi: So now the world is like someone who is suffering from malaria. It is heated, becomes hot, and gets malaria.

Emmanuel: Right, these climate changes we are getting is because the world is heated. Droughts come and we don't get rains when we expect, like these past ten years. It is caused by humans and humans should fix it.

Actor 1: So it is not God's fault?

Emmanuel: It is not God's fault. This is not caused by God. It is caused by humans themselves. Like those countries I mentioned. So don't cry to God, we are destroying our environment ourselves.

Actor 1: So humans are beating themselves?

Emmanuel: Yes, we are. It is caused by humans and humans should fix it. Yes, humans should have to stop cutting the rainforests we still have. And protect the world's forests for our future generation. Our protected trees will help meet the growing generation.

Fig. 5.2 Image from an educational video in which a documentary filmmaker explains to several Maasai from Terrat about CO$_2$ and industries as the causes of climate change (ResourceAfrica UK). "The World Has Malaria—ClimateConscious", 7:30, posted online by Max Thabiso Edkins, *YouTube*, 15 August 2012, https://www.youtube.com/watch?v=_Ay1snatTl8&t=2s

While this scripted video was created partly for educational purposes, considering its limited grassroots outreach possibilities, it may also have been intended to target donors and other policy-makers.[6] The clip offers vivid insights into the process of translating climate change— including the mediating forms, materiality of communication tools and practices—at the very end of the "translation chain". It reveals a glimpse of the climate narrative in the making; here, a story (supposedly) told by the Maasai to the Maasai, yet a clear NGO rhetoric can be discerned, in which a new allegiance is forged between Terrat and the globe. The actors are placed in the role of ignorant and helpless victims who need to be enlightened by NGOs, who operate as true "pedagogues of progress" (Englund 2011). At the same time, the villagers' "observations" are used to confirm that climate change is real and actually happening. As historical geographer Bravo (2009) has pointed out, Indigenous people around the world are used as "proxies" of climate change, warning us of a looming catastrophe that one day will befall all of us (cf. Farbotko and Lazrus 2012; Farbotko

6   The educational movie was shown at the COP 17 in Durban, where it won a prize. Produced by ResourceAfrica UK, in collaboration with Ujamaa Community Resource Team (UCRT) and Tanzania Natural Resources Forum (TNRF).

2010). The video implies that the Maasai, while wandering around, are complaining to each other about a hotter world, questioning what is going on in the world, and therefore seek recourse to the NGO for clarification. Of course, the reverse is actually happening: the NGO aims to enlighten the villagers, and makes them question "what all this is", prior to, and as a necessary step towards, their own intervention. The NGO is engaged in a process of "problematization": it defines the nature of the problem and then "makes itself indispensable" to solving it (cf. Rottenburg 2005, 2002; Roe 1999; Roe 1991; Ferguson 1990; Callon 1984).[7]

This fragment can be read as the (staged) culmination point of a positive-feedback cycle in which different epistemologies are brought together and (unconsciously) strengthen each other's gaze (see De Wit et al. 2018). It also reveals how the global is localized (e.g., the analogy with malaria, and how remote industries have a direct impact in Maasailand), and how the local is made global (plant trees to protect the globe). And while the polluting countries are named as the major culprit, the Maasai are also reminded of their own role as environmental destroyers. While I will touch here on the notion of cultural decline as trajectory narrative (see Rudiak-Gould 2013), inherent to the Maasai society's self-critique, I will not elaborate on questions of externally imposed blame and responsibility (i.e. blaming the victims), since I have written about this issue in another context in which it was more prominent (De Wit 2015).[8] Here, I focus on the

---

7   I have addressed this field of post-colonial power relations elsewhere, which pays more attention to the politics and practice of development and emerging frictions and shifting positionalities as actors move across different constituencies. In this development space, climate change becomes a political resource to maneuver oppressive politics, to leverage change and create opportunities in the absence of a benevolent state (De Wit 2019, 2018). These insights build on a larger body of literature of "brokers and translators" and critiques of development practice (Mosse and Lewis 2006; Mosse 2005; Bierschenk et al. 2002; Bierschenk et al. 2000; Rottenburg 2005, 2002; Roe 1999:5; Ferguson 1990)

8   See also the work of anthropologists Rudiak-Gould (2013) and Eguavoen (2013) on trajectories of blame and the translation of climate change. Their studies also demonstrate how climate-change discourses and local observations become part of the cultural narrative of "in-group blame". However, they come to diverging conclusions. Eguavoen sees self-blame as something that "ignorant" societies need to be liberated from through education, while Rudiak-Gould argues that we need to approach it as a form of local agency. The latter conclusion is in line with my own research findings. Yet an important difference is that in Rudiak-Gould's analysis

cultural and political meanings that emerge in the process of translating climate change through trajectory narratives that have been shaped by the complex and historically produced power structures within which this new discourse travels and is sustained. Trajectory narratives are defined by Rudiak-Gould (2013) as "discourses of the moral direction of society or the cosmos, with an associated sense of responsibility or blame for that trend" (Rudiak-Gould 2013: 10).

It is very tempting to understand the worldview of the Maasai (and other remote societies) in relation to natural dangers as "traditional", and radically different from a "modern" understanding of the world. As one of the binary fruits of enlightenment thinking, we too easily create a dichotomy in which societies that are without science and technology are driven by superstition, whereas modern societies are allegedly intellectually free and have an "objective" view of natural dangers, such as climate change. Following Douglas (1985; Douglas and Wildavsky 1982; Douglas 1966) the common view has been that for all other societies that have come "before us", every natural disaster is weighted with meaning and nature is thus seen as constructed and endowed with political, social, and moral significance. This is often thought to contrast with our understanding of modern society—that owes much to science—which is supposed to see nature as "it really is", morally neutral and politically empty (Douglas and Wildavsky 1982). The authors' objection to this alluring opposition is:

> But no! Try not to get into an argument about reality and illusion when talking about physical dangers. [...] On this subject we shall see that there is not much difference between modern times and ages past. They politicized nature by inventing mysterious connections between moral transgressions and natural disasters as well as by their selection among dangers. We moderns can do a lot of politicising merely by our selection of dangers (Douglas and Wildavsky 1982).

Building on "cultural theory of risk",[9] in the following I explore the role of culture in risk perception and how beliefs about (the interaction

---

local communities embraced climate change, while in Terrat people were overall much more reluctant towards accepting it.

9   In line with Rudiak-Gould 2013, I adhere to the "general" model of cultural theory of risk rather than Douglas's more rigid scheme based on group/grid distinction

between) society and nature, or worldviews, emerge as expressions of cultural priorities and preferences (Douglas 1985; Douglas and Wildavsky 1982) when the Maasai make sense of this new global threat. Following Douglas (1992), first of all (environmental) risks are undoubtedly always selected in ways that inhere in, protect and maintain a certain existing social order. Second, an important insight that has been reiterated by Rayner, Hulme and others in relation to climate change, is the universal phenomenon that moral and political beliefs are justified with appeals to nature (Hulme 2009; Rayner 2003). Nature is a mirror of societal behavior and thus operates as "a direct source of moral feedback for behavior, desirable or undesirable" (Rayner 2003: 278). The video fragment above underpins the highly-moralized tone of climate-change narratives. God is discarded as a causal agent and even fully removed from the explanatory horizon, yet anthropogenic causality and morality take center stage: "Humans are causing it, so humans should fix it".

To explain different forms of causality attributed to natural dangers, Douglas and Wildavsky (1982) distinguish between two forms of pollution. The first concerns a strict technical type (like river or air pollution) that rests upon a clear idea of a pre-polluted situation that can be measured precisely and carries no moral load. The second notion of pollution is nontechnical and indicates a contagious and harmful state, with mysterious origins, which carries the idea of moral defect (Douglas and Wildavsky 1982). The latter, nontechnical pollution, which they refer to as "pollution beliefs" or "pollution ideas", is of particular importance to my analysis. Yet when talking about observable changes in the environment or the weather, many of my informants referred to the first (technical) form of pollution. This is perhaps not surprising, as dwindling resources like grasses, plants, trees, rivers and water are more tangible, and indeed observable and measurable, than an abstract phenomenon like the climate. In the following analysis, I have reduced observations of climate change (or rather variability) to two main variables—rain and drought—because my respondents named them as the quintessential features of the

---

(Rudiak-Gould 2013: 9).

climate (rising temperatures were only rarely mentioned) in this environment.[10]

Nothing New under the Sun: On the Climate's Inherent Variability in Simanjiro

In a good year, the period of long rains produces an abundance of lush pastures, blossoming flowers, and sufficient crops like maize or beans to feed the family. It is a time of exceptional beauty and plenty. The rains give both people and cattle the chance to recover from the harshness of the dry season and regain their strength in preparation for the next dry season. There is no need for long-distance movement during this season, because the grasses and water can usually be found nearby. Families are united; there is time for leisure and ceremonies as there is enough milk and meat to share.

The contrast with the dry season could not be sharper: the area turns into a semi-desert with dust clouds covering the pastures. The water resources slowly dry up, and food for people and cattle gradually diminish. Both must travel greater distances to search for green pastures and water sources. The selected *illmuran* (young men, often called warriors) are sent out on their exploration journey (*eleenore*), and then report their findings in the traditional meeting (*engigwana engishu*) in which strategies for the dry season are planned. It is the time of scarcity. Women's daily activities become an ever-greater burden as women need to walk great distances to find firewood and water. Cattle and people lose weight and strength, and their concerns grow until the first signs of rain appear.

---

10   Although the climate consists of many more components than rainfall, its vital role in shaping and maintaining pastoralists' livelihoods (and religious convictions) legitimizes the overall focus on rain, and drought, as its counterpart (see Chapter 7, Attribution Science).

Fig. 5.3 Children milking a goat (photo by author, 2013), CC-BY-4.0.

The seasonal variations in Terrat's climate impinge powerfully upon peoples' day-to-day activities and their physical and social well-being, as well as upon the environment and the ways they perceive and relate to it. Mainly elderly villagers did not give testimony to (conspicuous) changes in the weather and the climate, nor were they familiar with climate-change discourse. Because they had never heard the term *climate change* (in any language), their accounts principally relied on the experience of sensory observations and (social) memory—i.e. the entire body of customary climate knowledge and environmental lore (Oba 2014). The ways in which social memory is passed on and reproduced in Terrat include warrior songs, in which *illmuran* sing about past hardships and adaptation strategies, and the names that are given to weather and social eventful years.

During interviews, my research partners and I asked informants whether they had observed changes in the weather, rainfall, or seasons compared to when they were young.[11] Their social memory echoed many dry spells, climate fluctuations and variability, years of hardships, environmental catastrophes, hunger, and events that were remembered

---

11   For translation of the concepts I relied heavily on the skills of my research partner Saruni Shuaka Kaleya, who did an extraordinary job of communicating the nuances in Swahili and Maa. Musa Kameika and Naini Morell were also indispensable in the whole research process. Language is crucial in the interpretation process, and especially challenging in this case as the term climate change is translated from Swahili to Maa, for which no official term is established, yet its linguistic politics reveal the struggles for meaning making. For a more detailed discussion of language and the translation dilemma, see De Wit 2017: 157.

by their own parents. Some elders recalled the years when they suffered from hunger. For instance, they mentioned years in which they were forced to eat the skin of their animals to survive. This was during the time when they were nomadic and did not use maize or other crops to supplement their diet. They pointed out two crucial coping strategies that enabled them to survive the dry season: mobility and drinking animal blood. Many elders referred to the period in which drinking blood was still a common practice, some with nostalgia but others with God-fearing disdain, which is a more recent sentiment inspired by Christian churches, mainly propagated by the Pentecostal church but is found in other denominations as well (De Wit 2017). But there was a consensus regarding the importance of both practices as vital survival strategies in the past. In brief, flexible adaptation mechanisms formed the lifeblood of a nomadic way of life in Maasailand, as hardships, environmental hazards, and catastrophes have been part and parcel of the highly variable climate.

Against this background, the first question related to pollution beliefs arises. If pollution indeed denotes an abnormal state, how does one, under such stochastic conditions, define the "normal" condition that is breached? In order to understand the reasons why climate change as a new global discourse enters as an alien idea, and is somewhat at odds with local realities, we need to make sense of the climate in Terrat and in the wider Simanjiro Plains, which is spatially and temporally highly variable (Leslie and McCabe 2013). This semi-arid environment is characterized by unpredictable, pronounced climate variability. As one informant explained, "[...] in our locality the climate knows a lot of fluctuations. One year you might expect rain and there will be no rain, in another year you expect drought and there is enough rainfall". Some informants found my question about changes in the rainfall compared to the past incomprehensible and referred to this variability as being the norm. For example, one responded: "of course the rains have changed; they have never been the same in this locality". A woman of approximately one-hundred years old reported, "There are no changes. The weather was like this from the very beginning. Sometimes there is less rain, sometimes there is more". The climate follows a bi-modal rainfall pattern, consisting of short rains (November—December) and a long rainy season (March—May/June). The fact that many people named different periods for the long and short rains also attests to an inherently variable climate.

Moreover, my informants pointed out that the unpredictable climate of this area was a great obstacle to cultivation, which was characterized as random trial and error. For example, one man told me that, "farming is some sort of gambling game, sometimes you win but more often you lose". Yet, despite frequent crop failures many villagers continue to cultivate, as the benefits of one successful harvest may outweigh the costs of many failed attempts.

Socio-Environmental Context of Simanjiro

Depending on the altitude, rainfall in the semi-arid rangelands of Simanjiro averages around 500–800 mm (Igoe and Brockingtion 1999) or 650–700 mm annually (Sachedina and Pippa 2009). This drought-prone region is among the most diverse and complex grassland savannah ecosystems in the world (Baird 2014; Olson and Dinerstein 1998). Despite the lack of meteorological data available for the wider area, the defining characteristics of unpredictability and fluctuation stand out. My informants highlighted the extreme spatial rainfall variability in the area. For instance, Meshack, the traditional leader, said, "It can rain here in this *boma*, but look at that *boma* over there and you see, there is no rain at all!" So how does one define a norm—and, consequently, a deviation from the norm—within such a hazard-prone and unstable environment? Despite the fluctuations, there is of course an approximation of regular seasonal patterns within the irregularity. Yet, while a clear dry season can be discerned, the rains differ greatly from year to year and place to place. Hence, some of the climate-cognizant informants found this notion of climate change challenging; for them it rather echoed something that the Maasai have already experienced in the past, as the following interview fragment with the traditional ruler of Terrat shows:

> Traditional ruler: The first time I heard about climate change was in 2009, when some people who were sent by the government were going around telling about climate change. And also there were some people who came to observe climate change [researchers] and I was taken to ewas (dry season reserve). I tried to ask them many questions, and they told me that it is the industries destroying the ozone layer. That is what these white men told me. [...] So, when I heard about these changes, I went

to Dar es Salaam, talking about climate change [together with Leboi]. [...] So now I really observed the changes because the Maasai have their own way of predicting rainfall, but nowadays it is very complicated. It is difficult to apply. You may apply but the answer may come in the wrong, wrong direction.

Researcher: And what did you think when you heard about it for the first time?

Traditional ruler: So, when I heard about this climate change for the first time, I thought it is just a repetition of the past. Because many years ago you may find five good years and then again seven not-so-good years. So, this is what I thought when I heard about climate change, that it is about these old changes of five good years, and seven bad years. But by then, I did not know about the industries yet. Because I am saying only what I know. Here in Tanzania there are no big industries, which could maybe destroy the ozone layer or change the climate of the earth.

Researcher: Do you understand now what it means?

Traditional ruler: I don't understand it very well because I did not carry on with those researchers. The government is not serving to educate people here in the village about climate change anymore. But I think that these changes are maybe those changes of five good years and seven not very much good years. We had many drought spells in our history.

[...] it seems that with our generation the situation is going in a bad direction. Also, the life expectancy of the people. In the past, people could live for 100 years up till 150 years sometimes. But nowadays a person can live only 70 years and we say "the man is too old, he is 70 years".

Notwithstanding the manifold statements about the lack of rain and intensifying drought and irregularities, communicating climate change is fraught with translation challenges, since global discourses on climate change tend to portray an unstable climate as a deviation from normality, albeit on much longer timescales than is recounted in oral memory. We spoke with an eighty-five-year-old man who had never heard of climate change. When we tried to explain it to him, he recalled the droughts of the past, and the years in which the Maasai suffered from hunger: "These are the same changes that we had in the past. First, there are seven good years, and then three bad years". His comments highlight how talking about changes in the weather and rainfall patterns cannot be detached from societal changes, since they go hand in hand with accounts of people's general wellbeing, altered lifestyle and customs, and eroding values and morals. This finding is in line with Fleming and Jankovic (2011), who advocate a less reductionist view of the climate as "a framing device in which the verities of life such as food, health, wars, housing, economy, social movement, or local identity change synchronically with *Klima*".[12]

It is crucial to bear in mind that the villagers of Terrat have abandoned their nomadic lives based solely on pastoralism, and cultivation and living in more permanent settlements is common nowadays due to a history of enforced relocations into permanent settlements and land alienation.[13] This has inevitably changed peoples' understanding of (and relation to) the climate. Mobility restrictions and the loss of grazing areas have changed their livelihoods and encouraged diversification strategies such as cultivation, and have deeply altered what it means to be Maasai. A middle-aged man commented on how shifting human— environment relations and changing perceptions of the climate go hand in hand: "It could be that the rains are less nowadays, but it is hard to tell because we are also more dependent on rain because we live and eat differently. Our mentality has changed".

In a similar vein, a middle-aged woman explained how she perceived a change in relation to ritual prayer, as opposed to the church:

> The years have changed, and people have changed. Nowadays people are eating a lot. In the past, you could have one bag of maize, but now you buy a lot of bags and put them in store, and you finish all of them. Families are big now. Now we also don't take blood because the church has prohibited the drinking of blood. And when there was no rain, we would just take a goat and women prayed to God and sacrificed a goat, then it would rain again. During the days of sacrifice it rained a lot.

---

12 Klima is of Greek origin. It refers to the inclination of the sun in relation to the earth, which partly regulates the global climate.

13 The Tanzanian government's attempt to relocate and "promote" agriculture and the sedentary life goes back to colonial times when the Maasai were forced to live in one of the most desolate areas in the country, but was also part of Ujamaa, which established a nationwide villagization programme that forcibly relocated people into so-called Ujamaa villages. In northern Tanzania, this resettlement was known as Operation Embarnat (Maa: "permanent settlement"), and many Maasai herders saw this operation as another attempt by the state to appropriate pastoral rangelands (Sachedina 2008: 110).

A lady, approximately ninety years old, spoke about land use and the difference in individual property as opposed to open access: 'There was one year without rain when I was a child that I cannot forget. Nowadays it happens more often. But in the past land didn't belong to anybody, so we moved until we found grass. Now everybody has his or her own land, so you have to stay in your own land.' Another very old woman remarked about changes in the weather: 'Perhaps the weather has changed, but we have changed too. We used to follow the clouds; nowadays we have settled.'

Thus, while following the clouds guided the Maasai's relationship with their environment and the climate, due to drastic changes in their ways of living, it is not fruitful to disentangle climatic and societal changes. What can we learn from these stories about change in relation to pollution beliefs? Following Douglas and Wildavsky (1982), since pollution beliefs are based on the idea of moral defect, they serve as a useful political argument because they trace causal chains from actions to disasters. Furthermore, as discussed below, these ideas about pollution are also the product of an ongoing political debate about the ideal society (Douglas and Wildavsky 1982). Talking about the climate is more than just a discussion of an atmospheric reality; it encompasses the socio-cultural, political and religious realms of society.[14]

On the Climate—Morality—Religion Nexus

Thinking about nature and climate in relation to culture and human society is as old as anthropology itself (and goes all the way back to intellectual forebears, who developed "climate theories", such as Hippocrates, Montesquieu, Ratzel, etc.; see Dove 2014). Recently, a large body of anthropological and social science literature is emerging that casts new light on these questions of morality, causality (blame/responsibility), and belief (rejection) in relation to climate change. Reception studies is one of such subfields (see Rudiak-Gould 2011) that seeks to account for a "located hermeneutics" (Livingstone 2005) by exploring the widely

---

14   This chapter focuses largely on the basic cosmology of the Maasai and the encounter with climate-change. Yet, I emphasize elsewhere that adaptation to climate change is intricately bound to the politics of land use and long history of marginalisation (de Wit 2018; 2017).

varying and radically different interpretations of the idea of climate change across the world. In the course of my fieldwork it became increasingly clear to me that what I perceived to be a climate change "void", in the sense that climate change was not prominently discussed, in fact revealed something very crucial, perhaps even existential. While I was interested in peoples' understanding of climate change and related (weather) observations, our conversations largely prompted stories about the loss of culture and respect, society, love, money, globalization, church, traditional religion versus Christianity—almost everything that encompasses the changing Maasai culture and ways of being and believing. Therefore, making sense of local climate accounts required broadening my "environmental horizon" and being attentive to the life worlds, practices and ontologies that overlapped in peoples' rain tales. For example, many elderly informants recalled that they and their forefathers had adapted to the climate by "following the clouds as long as we can remember" and deplored its current impossibility. And while pondering the climate and environment of the past, their nomadic lifestyle was often invoked, but so was their cultural identity and what it once meant to be a "true" Maasai. I argue that understanding these stories of cultural and environmental decline is vital for making sense of how climate change is interpreted and translated. As many scholars have demonstrated, these "prior commitments" (Jasanoff 2010) or pre-existing political norms (Cass 2007) profoundly influence how people assess, translate, respond to and render anthropogenic climate change meaningful (or not). Moreover, they shape the ways in which responsibility and blame are attributed to society (Eguavoen 2013; Rudiak-Gould 2013).

On Temporality and Degradation

While the older group of informants often emphasized the physical hardships of the past and did not romanticize the nomadic way of life *per se*, their social memory similarly spoke of ecological abundance, and echoed the vigor of cultural norms and values that they perceived to be prevalent in those days. Sufficient grasses, abundant trees, plenty of water, animals, and flowing streams were all imagined as part of the ideal past and mentioned on a par with the rootedness of cultural values such as respect, love, reciprocity, faithfulness and solidarity. In early ethnographies it is clear that mutual respect, *enkanyit,* formed the heart of Maasai culture and sociality. *Enkanyit* has been the lifeblood that sustained relationships of reciprocity and access to shared natural resources, and was expressed through appropriate greetings, forms of address and behavioral codes that differed according to age, gender, and kinship relations (Hodgson 2005; Waller 1999; Spear and Waller 1993a). The themes of love and respect were very prominent during interviews and referred to principles of solidarity and reciprocity that are said to be disappearing from society. Male informants often attributed this moral decline to encounters with other cultures, and processes of globalization and modernization, as Petro explained:

> People were spreading love a lot. So they were visiting each other during the night and during the day, to exchange ideas and to spread love. But this is now disappearing. These things are changing because of the introduction of other cultures.

A very old man related the changes of culture and rain in society to the introduction of industrial oil, which can be understood as a symbol of a monetary economy:

> Rain is a big problem. Things have totally changed and the one who changes these things is God. A big problem is lack of rain and human diseases. Increasing of diseases and this is due to industry oil that nowadays people prefer rather than butter. And also during the dry season—during hunger year—we drank blood from the cow and cooked it for kids, but nowadays you cannot do something like that. [...]

Men by and large spoke about money, modernization, and globalization in general, or a change in *oregie* (culture and customs). Fascinatingly, the group of elder women unequivocally deplored the eroding of *enkanyit* and *enyorotto*—the disappearance of respect and love—from society, which was often brought up in relation to the lack of rainfall. This female preoccupation with respect can be explained by a gendered conception of morality, as historically women were considered to be naturally more religious and closer to *Eng'ai* (see below). As such, they saw it as their responsibility to ensure the moral order of the daily world (Hodgson 2005); a role that they continue to assume and fulfil today (Hodgson 2011; see also the work of Todd Sanders (2004) on gender and reproduction, power, ritual and rain in Tanzania). A widely shared explanation given by female informants of the decreased rainfall was based on the moral conduct of society, and the lack of *enkanyit*. The following quotes from female informants illustrate the close association between rain, morality and (more implicitly) gender:

> The rains are bad nowadays because we are behaving badly; nobody is respecting each other any longer.

> People nowadays cannot even greet you properly any longer. Our society is losing its culture, and people do not love each other anymore. This is what is causing climate change to happen.

> When God decided to give us no rain, it is because we have sinned. People only want money nowadays. So they kill each other, have abortions and go to the mines.

> When I hear that there are changes in the weather this is maybe because of wrongdoings like homosexuality.

Attributing responsibility for natural disasters (or climate and weather anomalies) to society is a common way to protect a particular set of values, which belong to a particular way of life (Douglas and Wildavsky 1982). Sufficient rainfall denotes a condition in which a stable and peaceful world is imagined, and things are as they ought to be. God is content and expresses Her satisfaction by blessing people with rainfall.[15] In a year of severe drought, pollution beliefs speak about chaos and instability due to society's moral defects and invoke Maasai's cultural loyalty and religious faithfulness of the past. Faced with past and ongoing pressures from the forces of "modernization" and political economy, it is not surprising that for the Maasai, values that deserve protection are mostly related to their cultural identity, which they feel is under threat. In other words, discussions about the climate and the environment also have a temporal dimension in which a somewhat utopian and timeless culture (including knowledge and power) of the past is remembered, mirroring the ecologies of abundance.[16] As mentioned above, this does

---

15 The Maasai supreme being Eng'ai is almost universally referred to as female.

16 An anonymous reviewer commented that "the utopian vision of the Maasai past is not just an imaginary time of abundance; it's an account of nineteenth-century Maasai dominance of the regional political economy". My answer to this is that social memory is shaped both by historical events and trajectory narratives (i.e. a

not mean that the climate or environment were (imagined as) stable; rather, these trajectory narratives of the past should be understood in the context of conceptions of time about the future to which I now turn.

Two questions arise in this context. First, what can we learn from an imagined pristine past and a perceived-to-be eroding present? Second, how do we give meaning to a relationship between nature and morality that is inextricably intertwined? I first briefly explore the temporal dimension before probing the question of morality and the climate in more detail. As mentioned before, rain tales about the past are a commentary on an ideal present. To paraphrase anthropologist van Beek (2000, 1999), there is a strong identification between society and the climate, as the latter forms part of a shared identity. What follows from this observation is that "we are the climate", and conversations about the climate are meta-commentaries on society. As such, van Beek (2000) argues that climate-change discourses are a critique of society in which worries about the future are expressed. To explore the question of whether environmental degradation has repercussions on oral traditions (myths and other stories reflecting on the relationship between humans with their source of existence), van Beek explores projected futures in various cultures, and interrogates tales about end times.

Furthermore, van Beek argues that inherent to this message of decay, in which the present is less powerful than the past and the future is again less powerful than the present, is the notion of diminished resources as well as the loss of power and knowledge.[17] And based on my own findings, because relations of reciprocity are vital in times of duress such as drought, talking about the climate of the past similarly evoked a nostalgia for a time in which faith, culture, love and respect were strongly anchored in the social and cultural fabric of life. Finally, according to van Beek, the explanation of this sense of gradual loss in tradition and ecology over time stems from the dynamics of orality. So in rituals and local discourse, tradition is considered to encompass the wisdom and practices that are passed on from previous generations,

---

culturalist narrative of decline, see Rudiak-Gould 2014) that inhere in a society/ culture, and in which a particular future trend is imagined.

17   The discourses on tradition are different from the dynamics of tradition itself, which are flexible and adaptive (van Beek 2000). Horton also states that many African people see their cosmologies as timeless, which does not mean they are frozen (Horton 1975).

which is seen as a weak reflection of the past and leads to an even bleaker future (van Beek 2000). This observation indeed resonates with the ways in which my informants spoke about the current generation, and the fears they expressed about for example, "the direction of the bad changes for the future". Rain was a tangible trope for contemplation and what Rudiak-Gould calls "in-group blame" (Rudiak-Gould 2014), as the following quote from Rehema, a female elder, illustrates:

> Our culture is changing a lot because when we were young the children used to have respect for their parents. But this generation does not listen to their parents. [...] Today there is no love among the people; that is why life has changed. Today you can identify somebody who is poor, but in the past you could not because of love among people. [...] There are also changes in rainfall. In the past there was enough rainfall but currently there is not enough rain, which puts people in trouble. These changes in rainfall are caused by the church. Because in the past we used to sing in case of a bad year. We sang, gathered together with the women and sang to God "give us the rain". We were also going to *Oloiboni*. He told us "do this and do that" and the rain came. But today people are not visiting *Oloiboni* any longer—that is why it is not raining as usual.

Here too, temporality and the trajectory narratives of culturalist decline play a role, which go back to the colonial encounter but also to the more recent establishment of the first Christian churches. In the quote above, Rehema attributes society's moral failure and the lack of rain to the introduction of the church, although she is a Christian herself. When I asked her about this, she explained that early Christians were very different from the contemporary Christians, because the first churches (e.g., Catholic) were much more tolerant towards Maasai culture, while the more recently established denominations (e.g., Pentecostal) have suppressed varying forms of Maasai cultural expressions and material culture, like wearing ritual attires and jewelry. The trajectory narrative of cultural decline is coupled with in-group blame—a causal and moral explanation of that trend—in which the Maasai are ultimately to be blamed for their own cultural abandonment (cf. Rudiak-Gould 2013, 2014).

Her account also touches on other dimensions related to the second question posed above. In order to make sense of climate change "realities" (both observation and reception) in Terrat, we need to gain insight into the complex interconnection between the climate, morality and religiosity, which is in turn vested in local institutions and systems of power. It is important to note that, although we owe many conceptual polarities (e.g. nature/culture; body/mind; fact/value; subject/object) to Enlightenment thought and practice, the contrast between the "sacred", "religious", and "spiritual" as opposed to the "secular" realm is a residue of the same purification process (Armstrong 2009; Taylor 2007; Hodgson 2005; Latour 1993). Nevertheless, as Hodgson (2005) argues, while this opposition might hold true for some societies, the case of the Maasai reveals that until the twentieth century the secular and the sacred were one and the same.[18] Indeed, the Maa religion has no word for religion; they have adopted the Swahili term *dini* (from Arabic) (Hodgson 2005). Therefore, in line with Hodgson I use "religion" as an analytic term to probe and discuss the beliefs, convictions, (ritual) practices and experiences that express a relationship to some powerful divine being(s) or essence(s) (see Hodgson 2005), and to explore the stories that reflect on the Maasai's relationship to their source of existence (van Beek 2000).

On Continuity: *Eng'ai ai!*

> God is God. Who enables the sun to shine and the rains to fall. *Eng'ai* means three things: God, sky [or heaven] and all its contents, and rain (*Koko Theresia*).

The most common answers to the question of what could cause these changes in rainfall were "I don't know", "Only God knows", "We cannot know God's secret", "God brings us the rain", "God decides", "It is God's plan and She changed a lot", "Nobody can change the years but God and we can receive the changes from God", "Nobody but God is causing this, so nobody is causing it. Just the condition of the weather by God Herself is causing these changes". Thus, while the former section revealed how people are seeking an explanation within society—the climate as a mirror of society's behaviour—the realm of

---

18   Philosopher Charles Taylor, in his seminal book *A Secular Age*, has also stated that it does not make sense to distinguish between aspects of society as we do in the West, such as the political, economic, religious etc. Hence, the role of religion in "early" societies should be treated as interwoven with everything else rather than as a separate "sphere" of its own (Armstrong 2009; Taylor 2007: 2).

rain ultimately and exclusively belongs to the sole supreme entity, *Eng'ai*. And it should be emphasized that in addition to the gentle eschatology mentioned above—in which time might be running down and fading away gradually, an idea in which the present is a weak echo of the past—for the Maasai, the future remains an inaccessible domain shrouded in obscurity. Considering the precarious bond that has existed between Maasai pastoralists and rain, it comes as no surprise that the nature—culture connection is engrained with religious and symbolic meaning and ritual interaction. My conversations with people revealed that an intrinsic transcendental and moral connection exists between God, society, and the climate: indeed, the most significant word in Maa, *Eng'ai*, simultaneously means God, rain, and the sky (or heaven). The weather is a tangible manifestation of *Eng'ai*, a way to communicate Her satisfaction as well as discontent with the people. Sufficient rain is received as God's grace, while prolonged droughts are understood as a curse. Drawing on early ethnographies (such as Merker 1910; Hollis 1905; Johnston 1902), Hodgson (2005) describes the Maasai and their relationship to the environment and religious practices as follows:

> As pastoralists, Maasai had a close customary relationship to and dependence on the environment for their sustenance and social reproduction. Nature and its elements were understood as manifestations of Eng'ai or expressions of Her will, and were therefore central to Maasai religious beliefs and practices. The symbolic meanings of these aspects of nature were dynamic and contextual; they were shaped (and reshaped) through their use in ritual practice, and, in turn, shaped the form and content of these practices (Hodgson 2005: 25).

Although *Eng'ai* manifests Herself in different forms and goes by many different names, the Maasai religion is a monotheistic belief system. Despite the mass conversion to Christianity that took place in Maasailand relatively recently,[19] tales about the weather in relation to God and society form a testimony to this persistent and intricate relationship. While the Maasai's adoption of Christianity (or Christianity's adaptation to the Maasai) has entailed ruptures in certain cultural and religious forms and practices, a continuity in religious conviction and belief can be observed

---

19 For example, as late as 1969 it was estimated that there were no more than 300 practicing Maasai Christians in the Evangelical Lutheran Church, even though many more had been baptized (Groop 2006).

in exploring climate—society interactions. I argue that the continuity of an entrenched faith in *Eng'ai*, which helped smooth the conversion to Christianity, for the very same reason makes climate change (and the ways it is translated) difficult to accept.

The notion of an inner spirit or soul (*oltau*) is also important to the Maasai. As Hodgson demonstrates, conversion was seen as a process that took place outside of an individual's will; it was instead found in the readiness of a person's *oltau* (Hodgson 2005). Disentangling "Maasai religion" or cosmology from Christianity is unfruitful. As with any tradition or culture, Maasai culture, religious beliefs and practices have always been in flux, highly diverse and dynamic (Spear and Waller 1993b). Furthermore, what are now considered "Maasai" practices and beliefs are amalgamations that resulted from complex histories of exchange and earlier encounters with other ethnic groups. Christianity is also just as dynamic and idiosyncratic as the Maasai religion, and must thus be understood in its local expressions and manifestations (Hodgson 2005). For example, Terrat has a highly diverse proliferation of churches; the different denominations all have their own prescriptions, convictions, values, prayers, doctrine, (ritual) practices, (sacred) symbols, media, and material forms, etc. I do not seek here to map out the complexities associated with these various cults, but rather to point out that climate change is translated via the ontological and epistemological context that results from the encounter between Christianity and Maasai religion. While it is clearly impossible to exhaust all its constituents, we can distil some basic patterns of the pre-existing norms and commitments. As such, my take on Maasai "tradition" or culture aligns with a "coproduction theory", which treats it as the 'result of creative friction between past lifeways and newer influences' (Rudiak-Gould 2013: 11).

As an influential and externally imposed belief system that has travelled to Maasailand, we might very well draw lessons from the ways in which Christianity has taken root in the African context in general and in Maasailand in particular. Horton (1975) discusses comparative African pathways of conversion, and finds that the acceptance or rejection of Islam or Christianity is based less on the external religions themselves and more on the pre-existing thought patterns, values and socio-economic structure in the receiving society (what he refers to as their "basic cosmology"). However, instead of seeking to find selection principles in either the basic cosmology or in external variables, I focus on the interaction and relation between the two in the "fusion of horizons", which is the moment when an existential meaning occurs (Gadamer 1975). I approach interpretations of climate change as agentive cultural expressions or translations that emerge and become meaningful through existing interpretive horizons.

In the following, I discuss continuities and ruptures, with a focus on *Eng'ai* (supreme being) and the *oloiboni* (the local diviner and spiritual healer), who both appeared prominently in local rain tales.

According to my informants, the major difference between the Maasai religion and Christianity lies in the ways of worshipping—which relates to ritual practices and the enactment and strengthening of relations of reciprocity—and not in the "essence" of faith or believing in God itself. Informants spoke about faith at times as a way "to be faithful to" someone or something as a deliberate activity, thus implying a personal choice. There is a large body of literature on the anthropology of religion dealing with the difference between faith and belief, which is beyond the scope of this chapter, and I explore the terms here *ethnographically*. In their conversion process, their image of the divine being had remained largely unaltered; *Eng'ai* is still *Eng'ai*, as one elder woman explained:

> There are not two Gods, there is only a single God. The only difference is the way of worshipping. That is why people used to surround a tree called *oreteti*. That is a tree that they believed when you surround it and sing God can listen to your prayers. [...] And normally they went to *Oloiboni* and tried to make him their God. But actually, there is only one God.

The Supreme Being played (and continues to play) the quintessential role as provider of rain, and all-natural phenomena—particularly those related to the weather—were understood as a manifestation of Her divine powers. The weather thus echoed God's judgement—a common theme in religious traditions worldwide. In Maasai society, praying and asking for rain required mediation through authorized practices and ritual forms of purification, which were guided by the spiritual diviner and prophet.[20] Yet the realm of God (and thus the climate) could not be known. Anthropologist Spencer (2003: 6) has argued that the Maasai see God as an all-powerful and arbitrary force of providence whose intentions cannot be known, even by the prophets, "who have no more than a dim notion of the nature of the cosmos". As mentioned before, this notion of God (and the transcendental) as an undisclosed sphere inevitably influences ideas of the future. Spencer (2003: 6) goes further to argue that "the flamboyance of Maasai ceremony and self-regard is offset by a sense of resignation to an unknown and unknowable future. They see themselves rather like Plato's prisoners in a cave, ill-equipped to delve into ultimate truths".

While I am hesitant to speak of "ultimate truths", my informants highlighted the notion of an unknowable future; they refuted any attempt to probe future climate scenarios, or any ideas of the future for that matter. Questions about the future were always cast aside with laughter, followed by "we cannot know", or "only God knows". Yet mediation with God was common and necessary, as the Maasai used sacrifices and (communal) prayers in attempts to repent their sins, re-establish social and moral order, and reinforce their bond with *Eng'ai*. Women served as the midwives between *Eng'ai* and Her people, for women were given the power to create new life (Hodgson 2005; cf. Sanders 2004). The forms of mediation and intercession with *Eng'ai* have changed since Christianity has made its way to Terrat, which has affected ritual practices that for example regulated access to natural resources.

On Rupture: *Oloiboni*

In the past, the mediation of rain took place through the ritual prayer under a sacred tree (*oreteti*), or close to a water source by sacrificing a black sheep or a goat.[21] As women were considered to be more religious than men, to have a greater sense of spirituality and to be closer to *Eng'ai* (Hodgson 2005), this ritual was carried out by several women (in different stages of fertility) who prayed and sang naked while surrounding a dam. This ceremony was guided by the *oloiboni*, who was

---

20   According to my informants in Terrat, this practice has drastically declined over the years, but is still widely practiced in some other parts of Maasailand.
21   It must be a black one because this relates to morality and signifies a gentle God.

responsible for the timing and for giving instructions to the participants. The rain prayer now takes place in church, under the guidance of a pastor. Yet any member of the church (not only the pastor) can instigate the rain prayer. Whereas many elder women did not rejoice in this Christian democratic principle of praying for rain, others argued that the God of the church is still capable of bringing rain, "perhaps a little bit slower though" and "less powerful than in the past". Overall, there is a decreased sense of the bond between God, society, and rainfall, for which people sought an explanation in either the lack of ritual or the loss of faith.

Mediating rainfall has always occurred through the dominant (religious) authorities and was brought about in connection with a transcendental force, *Eng'ai*. This hierarchy inevitably touches upon notions of power, and the question emerges: who has the authority to mediate (the epistemics of) the climate and rain? Female informants in particular have attributed the lack of rain to the fact that this ritual prayer through sacrifice has ceased to exist and blame the church for prohibiting this practice. One older woman recalled the effectiveness and power of this prayer: 'After we carried out this practice by surrounding a dam, the rain occurred the same day and not any other day. The same day the rain fell down. But nowadays the church is not allowing us to go'.

Fig. 5.4 *Koko* (elder women, or grandmothers) dressed for a circumcision ceremony in the neighboring village Sukuro (photo by author, 2013), CC-BY-4.0.

The ritual prayer guided by the power of the *oloiboni* was important to bring rain, but faith also played a vital role. As one old man (non-Christian) put it:

> When I was young people went to pray for rain with *oloiboni*. Praying for rain is about faith. When people went they had very strong faith, they were very faithful to him. Nowadays people go to church because they love church. It is difficult to know if they are truly faithful. The *oloiboni* is not remembered because they pray for Jesus. But the difference was that people were faithful and he gave the instruction of what to do. Now in church there are no sacrifices. In church there is only preaching.

A middle-aged man who was quite happy in church explained: "We do not longer believe in the power of *oloiboni* so we directly pray to God. Sometimes we make a special prayer for rain in church so that the rain may come. God normally listens to our prayers. When we realized that the rains came when we prayed to God, we left the *oloiboni*".

However, many women made opposite claims. While generally being content with the church, they regretted that the ritual rain prayer—and their own role as moral mediators therein—no longer existed. While expressing their disdain about the contemporary *oloiboni*, in terms of channeling rainfall they found them to be good ritual leaders in the past, as they gave precise instructions about what to do. It is crucial to note that the *oloiboni* was not the one empowered to bring rain himself; rather, he was endowed with the power to mediate rain through *Eng'ai*. As one *koko* explained: "We visited the *oloiboni* and he tried to do something. Sometimes it would rain but not because of the *oloiboni* but only because God decided to give us rain". Rain and power are intrinsically connected, particularly in a context where rain sustains all forms of life (for studies on rain and power in Tanzania, see also Sheridan 2012; Sanders 2004). Therefore, it comes as no surprise that rain is seen as a sacred matter that in part, both conceptually and linguistically, coalesces with the Supreme Being. When it rains, people exclaim, "There is *Eng'ai!*". The disappearance of the ritual rain prayer—accompanied by the collective sacrificing of a black animal as a gift for *Eng'ai*—has entailed a weakening of the ways in which people (mainly women) sought redress for society's behavior and maintained a relationship with *Eng'ai*. This ritual power and redress were important regulatory principles that guided access to natural resources and relations of exchange and reciprocity, which were crucial in adaptation to climatic fluctuations and drought.

Hence, they also find the explanation for the lack of rainfall in the eroding power of the *oloiboni*. Apart from the church, there is a more intrinsic reason for the *oloiboni's* loss of ritual authority. When I asked people whether they still believed in the power of the *oloiboni*, many replied that this depends on the faith of the people. As Leboi explained: "God also depends on the faith of people. If somebody believes in the church he may succeed there, if they pray to the Maasai God they may also succeed there. It depends on the faith of the people".

Because of people's strong resentment of the diviner, I had assumed that this attitude also entailed a new and exclusive ontology that discarded his powers altogether. However, it appeared that the *oloiboni's* powers are not necessarily "unreal" or that they cease to exist, but rather that getting in touch with him or establishing a bond involves engaging with ungodly spirits and the realm of the "occult". Furthermore, just as one's spirit or soul (*oltau*) operates like another inside oneself, who decides for you what to believe, the power of the *oloiboni* is not an intrinsic agentive force that belongs exclusively to him that he can exert over the people. Put otherwise, in addition to his own derogatory practices (that have become known as such as propagated by the Christian church) that in part corroded his power, for the *oloiboni's* power to "work" he needed the dialectical enforcement from below that is enacted through people's faith. As one (Lutheran) man recalled: "When I was young others went to pray for rain with the *oloiboni*. It is about faith. [...] The difference was that people were very faithful, and he [*oloiboni*] gave directions for what to do in case of diseases. Now in church there are no sacrifices. In church, there is only preaching".

In this section I have explored the historically complex and intricate relationship between society, rain, and God, in which morality featured prominently and was sustained through ritual prayer and other ceremonies. These rituals fulfilled a particularly vital role in the face of severe drought in order to maintain social cohesion, strengthen relations of reciprocity and keep society together (for a comparison with Brazil, see Taddei 2013; 2012). Furthermore, whatever the external influences that brought about new cosmological configurations, and institutional or ritual ruptures, the deeply entrenched notion of *Eng'ai* remains the quintessential lifeblood that serves as the foundation for this bond.[22]

Climate Change as Disenchantment

The radio is one of the primary sources of information on global discourses in Terrat. The radio's reach in Maasailand makes it the only platform in their own language that weaves connections to international and national concerns and local environmental realities. Orkonerei Radio Service (ORS community radio) is headquartered in Terrat.[23] Its main vision is to enhance knowledge and educate the pastoralist community for the purposes of development by promoting accountability, equality, peace and unity. The service also seeks to empower Maasai women by establishing radio community groups and donating radios to women. However, I found that while the majority of men own a radio and listen regularly to radio programmes, women do so to a much lesser extent. Moreover, when women are at home, they have less time than men to sit and listen because they are burdened with an array of time-consuming household tasks. Hence, among my informants there were more climate-cognizant men than women, and often they had received the relevant information from the radio.

The ORS is an important source of information for the Maasai because it is the only radio station that broadcasts in the Maa language. It connects the Maasai to global concerns and mediates culture and tradition. In a sense it thus invokes and remediates notions of Maasai identity. The story of Babu (elder man) is a good example. When the radio station was established Babu was invited to present a programme

---

22   This is not to say that *Eng'ai* is stable and unchanging, as peoples' relation to Her were and still are continuously enacted and sustained through certain cultural and religious practices, but because both Christianity and Maasai religion are monotheistic belief systems, conversion afforded a certain continuity in religious experience of a continued existence of the Supreme Being. This is different from the encounter with Christianity and many other local/Indigenous religions in the world, where local deities and animist ontologies were largely perceived to form a threat to the gospel.

23   The radio station was initially owned by the Institute of the Orkonerei Pastoralist Advancement, which is a voluntary community membership organization founded in 1991. From 2009, the of the radio station's ownership was transferred to Orkonerei Mass Media, a Maasai-run company.

called "*Orngara le Maa*" [to unite the Maasai] about the Maasai language and culture. The programme seeks to unite all the Maasai throughout Tanzania and Kenya by informing people about their language and traditions. During the programme, Babu receives phone calls from Maasai from many different regions and they ask him questions about particular Maa words, ceremonies and habits, etc. According to Babu, this radio programme is very important because he has observed many cultural changes that are eroding important values such as *enkanyit*. He explained:

> People are not following the Maasai tradition as in the past. For example, to make holes in the ears. Also, when there was no milk they could drink the blood from the cow but nowadays people are no longer doing that. The major effect is that people are leaving the culture and some people are dying.

For Babu, the radio programme is very important for keeping Maasai traditions alive; reminding the Maasai of their own traditions and language. However, while remediating tradition and cultural values, the radio also plays into the complex relationship between Maasai traditions and the disenchantments associated with modernity. As Englund (2011: 2326) compellingly demonstrated in his study of the role of the radio (and the hundreds of mundane stories that radio listeners shared) in Malawi, "insights emerge into how Africans pursue their desires under the condition of globalization".[24]

But in Terrat, the radio also increases listeners' disenchantment with the world, and informs them of the dangers of globalization. For example, people hear about scientists who reject the existence of God, about nuclear wars, environmental destruction, industries, explosive weapons, air pollution, cancer, and climate change. When Babu heard about climate change on the radio, he thought it referred to a change in culture and the lack of respect in society. One ORS radio programme "*Mazungumzo ya Mazingira*" [Talk about the environment] devoted an entire episode to the topic of climate change. A special guest was invited, called Hassan, to explain the concept to the people:

---

24 The role of the radio in Maasailand merits its own ethnographic study. See for example the work of Englund 2011.

Hassan: It is the change based on environment, e.g. temperature, moving winds and rainfall. These changes are repeating all the time. [...] Scientific investigation found that changes are occurring [...]. Climate change can be human induced, and it is also naturally occurring. This is because human activities pollute the atmosphere. Human activities like burning charcoal and burning forests increase the level of CO_2 in the atmosphere and causes global warming. The industrial revolution, which began in 1730 caused emission of poisonous gases that made a layer of thick gases, and allows rays to penetrate, but prevented gases from escaping. These poisonous gases come from different places such as industries, volcanic eruption, burning charcoal or forest, explosion of weapons used in wars. [...] Trees that normally absorb carbon dioxide are being cut and this is very dangerous.

His advice to the people was to minimize industries and the burning of forests, and to motivate people to plant trees to reduce gases like CO_2. This explanation seems rather complicated for people who have never heard of CO_2, the atmosphere, the industrial revolution, or even science. Moreover, a giant leap is made between the global causes (industries) and local responsibility (planting trees). Another invited guest explained his ideas about climate change:

Frederik: This is caused directly by human activities. How does this climate change take place? Like I said. Directly by human activities. A human being because of her poverty can go into the forest and cut down trees for different purposes. For example, for charcoal burning and the smoke that is produced in this process is directly released into the atmosphere and causes climate change. This ozone layer is destroyed, and some holes are appearing, ozone layer depletion. [...] We are also the causes of this thing, since we are cutting many trees, burning charcoal [...].

The rest of the programme was dedicated to tree talk and how to sensitize people about tree planting and prevent them from cutting down trees. The anthropogenic approach to the climate features prominently in climate-change discourses. The NGO in the video fragment described above also explained to the group of actors that they should "stop praying to God for it is not God's fault", and that they have to plant trees instead, since humans—including the Maasai—are responsible. While in the educational movie the actors obviously accepted this scientific explanation, as Leboi has been perfectly able to reproduce it in different public fora (including in his accounts to me), the reality of interpreting and translating climate change proved much more subtle, stubborn and complicated. In one of my many conversations with Leboi, he asked me a question that I had grown familiar with in the course of my stay in Terrat, and it evoked a fascinating and controversial point in the translation of climate change: "Is it true that there are people in Europe who call themselves scientists and who don't believe in God?"

> Researcher: Yes, that is why many people in Europe maybe don't believe in God because they cannot find proof that God exists, because science is all about proving things and if you believe that you have to prove something before it can exist, you might lose faith in God. And instead you focus on science and technology. What do you think about science Leboi?
>
> Leboi: I do not know the meaning of science. Probably you learn about that when you get education.
>
> Researcher: Okay, well for example this climate change is based on scientific findings by measuring the temperature and rainfall.
>
> Leboi: If these scientists are saying that climate change is happening, I can say it is true, these men may be correct in some things maybe. But if those scientists are saying there is no God, they are wrong. Because when I was growing up my parents never told me that God was present. But one day when I was still very young I was walking in the forest. I was alone. Suddenly, a very big and dangerous snake appeared and then I screamed: "Eng'ai ai! Oh my God". I just mentioned it. So, I questioned myself and wondered: if my soul [oltau] itself seems to know about God, then really She must be present.
>
> Leboi: So if these scientists say that God does not exist, do they have a different God?

Leboi's confusion about the meaning of science was quite common; several others inquired about it too. And despite the scientific rationale that dominated these climate-change explanations, according to my informants it was impossible as well as completely senseless to reduce the narrative solely to a secular causality. The sheer idea of having no God (as those scientists and some NGO workers appear to claim) must mean that they have at least a different God, as Leboi's *oltau* has revealed to him. The ways in which climate-change discourse and its scientific underpinnings is translated by different mediators (radio makers, NGOs, educators, government officials, scientists, etc.) can be characterized as a hermetically sealed ontology in which human-induced climate change has discarded God as a causal agent or final moral arbiter. This explanation or translation does not resonate with the ways in which the villagers of Terrat perceive and relate to the climate, which is a cosmology based on inclusivity and continued transcendence. While for some it seemed possible to embrace (at least parts of) the new cosmological configuration called climate change—including a new vocabulary that speaks of industries, CO_2, and the atmosphere—nature's entanglements remain characterized by inclusivity, with *Eng'ai* as the seat of life and rain. This particular translation of climate change seeks to dislodge the weight of a relationship between rain, God and society, with a profane worldview emptied of God and purged of "superstition". The scientific complexity of the story also played a part in the demystification of nature. From Leboi's account we also learn that accepting or rejecting climate change is not a binary choice. His meaning-making process is agentive, as he put forward his own conditions within which he can accept (a part of) this new prophecy. In other words, he is willing to accept the explanation for the idea that the weather/climate is changing, but not if he cannot stay true to his own set of values.

The example of Rebecca, a middle-aged Maasai woman living in Terrat, exemplifies a more general point about causality. We met Rebecca in Arusha during one of the many climate change workshops in which my research partner and I participated. Several Maasai villagers attended this workshop from different regions along with a few researchers and NGO workers. During the meeting, which was designed to sensitize grassroots people to the subject of climate change, I wondered what residents of a village like Terrat would possibly make of these graphs, models and statistics presented to them in PowerPoint. After we realized that one of the participants lived in Terrat, we got the chance to ask her these questions once we were back home. She explained that she was invited by PINGOs, the organizing NGO, after attending a village meeting in Terrat. When we asked her what she thought about climate change she replied: "It was very tough to learn about climate change because it was my first time and it was very complicated. But I myself cannot know what causes these changes. Only God knows. Only God is the One who changes everything".

For Rebecca, *Eng'ai* remains axiomatic in the ultimate judgment of the world, manifested in the weather. Similar to the process of embracing Christianity, the new climate-change discourse involves several approaches to belief and understanding that are not necessarily mutually exclusive. Leboi contends that the realm of rain is under the purview of *Eng'ai* and cannot be explained by science alone: human beings or industries might play a role in destroying the environment, but it is up to God whether to endow people with rain or not. In a similar vein, villagers perceived the role of scientists (at least what they were told about science and scientists)—who not only claim to be able to predict the weather, but also discard the existence of God—as some sort of self-declared apotheosis, the ultimate form of hubris. For indeed, nobody but God can decide about the rain and the climate. One of the schoolteachers who measures rainfall in Terrat explained to us that "People do not trust the information from the weather forecast on the radio because they do not know where they get the information from. Because there is no communication with God. Only if they communicate with God will they trust it". A middle-aged man (here *mzee*, Swahili for older man) who seriously questioned the notion of climate change asked me the following revealing questions:

Mzee: We hear nowadays that scientists say there is no God, is it true?

Researcher: Yes. Some at least. [...]

Mzee: I think there is a God because She is the one who created us. Do you have an oloiboni in your country? [...]

Researcher: No. Not really [...]

Mzee: In your country there is no rainfall? Because in our country we say the rain comes from God. How do you explain rain? What about a thunderstorm? We say that it is God talking. What about in your country?

Researcher: We say that it is just friction between hot and cold air.

Mzee: Ai, ai, ai, it seems that science is really trying to say there is no God at all!

The role of the scientist is perceived as a self-declared "weather prophet" and is therefore rejected, for he assumes to be able to unearth God's secret and thus incorporate Her power. It appeared that some cognizant informants were (in their own ways) able to relate to the scientific rationale underpinning climate change and its anthropogenic origin, and to take it seriously. But the notion that science could develop models for the future was irreconcilable with their own worldview. Eli, a forty-year-old man, explained:

> I have heard about climate change on the radio. First of all, temperatures have increased. And also rainfall has declined. Industries which emit fumes in the atmosphere are causing this climate change to occur. And also, the application of weapons, explosive weapons during the war which turn the ground soil to come up. And also, cutting down of trees. [...] But nobody can answer the question how it will be in the future, because nobody can predict.

Although the church leader has taken over the ritual power to pray for rain from the *oloiboni*, the deification of the latter ultimately led to his own symbolic (and institutional) downgrading. Thus, whereas the pastor now assumes only a mediating role, the *oloiboni* has been taken to task for claiming to possess prophetic qualities, inspired by the Christian values that have been propagated. The fate of the scientist—who not only intervenes in the realm of *Eng'ai* by claiming to have forecasting skills, but at times even denies the existence of God altogether—seems to be similarly disqualified.

Yet the fact that people did not express "trust" in science and scientists does not mean they discarded all anthropogenic causes of climate change. As mentioned before, there has always been a strong bond between society, morals and nature, which means that humans *do* have agency to bring about cosmological harmony. From historical sources, we learn that *Eng'ai* (as in God, sky and rain) had a dialectical relationship of mutual dependency with *enkop* (the land or earth). So there was a complementarity between *Eng'ai* and *enkop*, heaven and earth; both possessed the agency to alter and maintain the relationship, for both humans and *Eng'ai* created and nurtured life (Hodgson 2005). So, bringing deforestation, population growth, and even industries and CO_2 into the causal chain of blame and pollution beliefs was for some informants acceptable and reconcilable with their own "prior commitments". But discarding *Eng'ai* as an agentive force was considered to be a true disenchantment of the world.

The traditional leader who was very well aware about climate-change discourse, contested the initial Maa translation proposed by the radio station workers, *engibelekenyata engijape engop* [a change of air and earth]. He proposed a remarkable alternative: *engibelekenyata* (a change by *Eng'ai*). He explained: "You talk about rainfall, plants, temperature. Is there anybody who can change these things? No! Only God!" He also contended that the official Swahili version should not be translated literally (*mabadiliko ya tabia nchi*). As he stated: "*Nchi* [meaning country, earth or ground] is referring to us, but it is *not* us. We should not try to translate directly, because it will be misleading. You should look at what this means for us". According to the traditional leader, while explaining that industries are the principle cause of climate change, the domain of rain, air, and sky (or heaven) is, ultimately, in the hands of *Eng'ai* (De Wit 2018).

Concluding Reflections

Recently, scholars have begun to critique attempts that view religion as the basis for climate denial/rejection and thus merely as barriers in the larger pursuit of adaptation. What these perspectives share is the idea of a somewhat passive recipient whose local belief system stands in the way of a correct interpretation of climate change (Fair 2018; Kempf 2017; Donner 2007). Instead, I argue against this purification and rather wish to foreground the need for what Fair calls, "the potential for more-than-scientific yet not anti-scientific responses that are locally meaningful and morally compelling" (Fair 2018: 11). More critical scholars have recently identified different explanations for denial, such as in-group blame and trajectory narratives of decline (Rudiak-Gould 2014; 2013), or the incommensurability between specific spatio/temporal configurations (Rubow and Bird 2016); or the extent to which it offends a separation between sky and earth (Donner 2007). According to Donner (2007), denial and skepticism are a philosophical problem and are due to the conflict between an idea of human-induced climate change on the one hand, and the ancient belief that earth and sky are two separate domains on the other (Donner 2007). Donner's point seems to hold true in Maasailand, yet there appears to be room for a "negotiated reading" (Hall 1980)[25] in which an interpretation becomes possible that is not necessarily antithetical to science. This means that, for example, both the industries and moral decay within society can be blamed for climate change, while at the same time it is in God's hand to express and define the directionality of this change. By moving away from the idea that the public has a "knowledge deficit", the emancipatory potential of public engagement approaches such as reception studies, lies in the fact that it sees the "plurality of perspectives among the lay public as the place to debate and build democratic consensus" (Bravo 2009: 4).

In this vein, I propose that the traditional leader's pronouncement that "You should look at what this [climate change] means for us, the Maasai" is the Maasai prerogative for translating climate change. I have argued that the overall lack of climate-change discourses can only partly be explained by an absent state that continues to neglect one of the country's most marginalized populations. As such, global climate narratives that foreground crisis scenarios do not resonate with local concerns, as people have other pressing issues to deal with: they lack the most basic access to health care, education, water, land, pastures, infrastructure, medication, vaccination and so on (De Wit 2018). In this chapter, I have argued that we need to examine the complex set of prior commitments—including entanglements, ruptures and continuities—that shape how this new narrative is appropriated, received and translated. I have sought to explore the historically produced norms of the Maasai villagers of Terrat within the broader historical, environmental and socio-cultural "interpretive context", as well as epistemological and ontological dimensions that account for the possible incompatibilities between (the translation of) global discourses and local realities. Against the background of the longstanding marginalization of the Maasai and increasing pressure from a globalizing world, (partial) rejection of climate change as an incipient new doomsday scenario about the world should not be understood as a form of ignorance, but

---

25 Literary theorist Stuart Hall, who is closely associated with reception studies, proposed a communication model of encoding and decoding as a form of literary analysis in media studies. He focused on the key role of the reader/viewer, exploring the possible degree of negotiation and opposition among the audience and distinguishing three positions: (1) dominant (hegemonic) code; (2) oppositional (counter-hegemonic) code; (3) negotiated code (Hall 1980).

rather as an emergent knowledge space that affords agentive potential for resistance. In line with cultural theory's basic commitments, risk perceptions emerge in ways that enable people to stay true to one's own set of values. Comparable to Rudiak-Gould's findings in the Marshall Islands, among the Maasai in Terrat a culturalist narrative of decline, embedded in a gentle eschatology, speaks lamentably about the Maasai's historical engagement with the forces of modernity (e.g. Hodgson 2011), leading to in-group blame about the loss of respect (*enkanyit*) in society. Yet, contrary to the Marshall Islanders who almost unanimously embrace the new prophecy, the Maasai picture appears to be more heterogeneous and fraught with friction and incommensurability. This means that prior commitments alone cannot account for or predict how interpretation (ultimately) takes place, but more attention is needed for a contextual or empirical hermeneutics that explores the dynamic appropriation process within which reading and explaining takes place. I have demonstrated that the Maasai's relationship to the climate (as an element of the complex entanglements between Christianity and Maasai cosmology) remains embedded in an inclusive ontology in which society, morals and nature are interwoven—a way of living that ceases to make sense when purged of *Eng'ai*. The relatively new field of reception studies in climate anthropology is only beginning to reveal its potential to understand the various contextual factors that shape how people interpret and give meaning to the contemporary idea of climate change.

References

Armstrong, Karen. 2009. *The Case for God* (Toronto: Random House of Canada)

Baird, Timothy D. 2014. "Conservation and Unscripted Development: Proximity to Park Associated with Development and Financial Diversity", *Ecology & Society*, 19.1: 4, https://doi.org/10.5751/es-06184-190104

Berry, Evan. 2016. "Social Science Perspectives on Religion and Climate Change", *Religious Studies Review*, 42.2: 77–85, https://doi.org/10.1111/rsr.12370

Bierschenk, Thomas, Jean-Pierre Chauveau, and Jean-Pierre Olivier de Sardan (eds). 2000. *Courtiers en développement: Les villages africains en quete de projets* (Paris: Karthala & APAD)

——. 2002. *Local Development Brokers in Africa. The Rise of a New Social Category* (Mainz: Johannes Gutenberg University)

Bravo, Michael T. 2009. "Voices from the Sea Ice: The Reception of Climate Impact Narratives", *Journal of Historical Geography*, 35.2: 256–78, https://doi.org/10.1016/j.jhg.2008.09.007

Callon, Michel. 1984. "Some Elements of a Sociology of Translation. Domestication of the Scallops and the Fishermen of St Brieuc Bay", *The Sociological Review* 32: 196–233, https://doi.org/10.1111/j.1467-954X.1984.tb00113.x

Cass, Loren R. 2007. "Measuring the Domestic Salience of International Environmental Norms. Climate Change Norms in American, German and British Climate Policy Debates", in *The Social Construction of Climate Change. Power, Knowledge, Norms, Discourses*, ed. by Mary E. Pattenger (London: Routledge), pp. 23–50

De Wit, Sara. 2015. *Global Warning. An Ethnography of the Encounter Between Global and Local Climate-Change Discourses in the Bamenda Grassfields* (Cameroon: Langaa RPCIG), https://doi.org/10.2307/j.ctvh9vvwx

——. 2017. "Love in the Times of Climate Change. How the Idea of Adaptation to Climate Change Travels to Northern Tanzania" (PhD thesis, University of Cologne)

——. 2018. "Victims or Masters of Adaptation? How the Idea of Adaptation to Climate Change Travels up and down to Maasailand, Northern Tanzania", *Sociologus*, 68.1: 21–40, https://doi.org/10.3790/soc.68.1.21

——. 2019. "To See or Not to See: On the 'Absence' of Climate Change (Discourse) in Maasailand, Northern Tanzania", in *Environmental Change and African Societies*, ed. by Ingo Haltermann and Julia Tischler (Leiden, Boston: Brill), pp. 21–47

De Wit, Sara, Arno Pascht, and Michaela Haug. 2018. "Translating Climate Change: Anthropology and the Travelling Idea of Climate Change", *Sociologus*, 68.1: 1–20, https://doi.org/10.3790/soc.68.1.1

Donner, Simon D. 2007. "Domain of the Gods: An Editorial Essay", *Climatic Change*, 85. 3–4: 231–36, https://doi.org/10.1007/s10584-007-9307-7

Douglas, Mary. 1966. *Purity and Danger. An Analysis of Concepts of Pollution and Taboo* (London, New York: Routledge).

——. 1985. *Risk Acceptability According to the Social Sciences* (New York: Russell Sage Foundation).

——. 1992. *Risk and Blame. Essays in Cultural Theory* (London, New York: Routledge).

Douglas, Mary, and Aaron Wildavsky. 1982. *Risk and Culture. An Essay on the Selection of Technological and Environmental Dangers*, 1st edn (Berkeley: University of California Press)

Dove, Michael. 2014. *The Anthropology of Climate Change. An Historical Reader*, Wiley Blackwell Anthologies in Social and Cultural Anthropology 18 (Hoboken, New Jersey: Wiley/Blackwell)

Dürr, Eveline, and Arno Pascht, Arno (eds). 2017. *Environmental Transformations and Cultural Responses. Ontologies, Discourses, and Practices in Oceania* (New York: Palgrave Macmillan), https://doi.org/10.1057/978-1-137-53349-4

Eguavoen, Irit. 2013. "Climate Change and Trajectories of Blame in Northern Ghana", *Anthropological Notebooks*, 19.1: 5–24

Englund, Harri. 2011. *Human Rights and African Airwaves: Mediating Equality on the Chichewa Radio* (Bloomington: Indiana University Press)

Fair, Hannah. 2018. "Three Stories of Noah: Navigating Religious Climate Change Narratives in the Pacific Island Region", *Geography and Environment*, 5.2: 1–15, https://doi.org/10.1002/geo2.68

Farbotko, Carol. 2010. "Wishful Sinking: Disappearing Islands, Climate Refugees and Cosmopolitan Experimentation", *Asia Pacific Viewpoint*, 51.1: 47–60, https://doi.org/10.1111/j.1467-8373.2010.001413.x

Farbotko, Carol, and Heather Lazrus. 2012. "The First Climate Refugees? Contesting Global Narratives of Climate Change in Tuvalu", *Global Environmental Change*, 22.2: 382–90, https://doi.org/10.1016/j.gloenvcha.2011.11.014

Ferguson, James. 1990. *The Anti-Politics Machine. "Development," Depoliticization, and Bureaucratic Power in Lesotho* (Minneapolis: University of Minnesota Press)

Fleming, James Rodger, and Vladimir Jankovic. 2011. "Revisiting Klima", *Osiris*, 26.1: 1–15, https://doi.org/10.1086/661262

Gadamer, Hans-Georg. 1975. *Truth and Method* (New York: Seabury Press)

Groop, Kim. 2006. *With the Gospel to Maasailand. Lutheran Mission Work among the Arusha and Maasai in Northern Tanzania 1904–1973*. Åbo: Åbo Akademis Förlag

Hall, Stuart. 1973. "Encoding and Decoding in the Television Discourse", Paper for the Council of Europe Colloquy on Training in the Critical Reading of Televisual Language, University of Leicester, September, https://www.birmingham.ac.uk/Documents/college-artslaw/history/cccs/stencilled-occasional-papers/1to8and11to24and38to48/SOP07.pdf

——. 1980. "Encoding/Decoding", in *Culture, Media, Language: Working Papers in Cultural Studies, 1972–79*, ed. by Stuart Hall, Dorothy Hobson, Andrew Lowe, and Paul Willis (London: Hutchinson), pp. 128-138

Haraway, Donna. 2016. "Staying with the Trouble. Anthropocene, Capitalocene, Chtulucene", *Anthropocene or Capitalocene? Nature, History, and the Crisis of Capitalism*, ed. by Jason W. Moore (Oakland: PM Press), pp. 34–76

Hastrup, Kirsten. 2015. "Comparing Climate Worlds: Theorising across Ethnographic Fields", in *Grounding Global Climate Change. Contributions from the Social and Cultural Sciences*, ed. by Heike Greschke and Julia Tischler (Dordrecht: Springer), pp. 139–54, https://doi.org/10.1007/978-94-017-9322-3_8

Hermann, Elfriede, and Wolfgang Kempf. 2018. "'Prophecy from the Past': Climate-change discourse, Song Culture and Emotions in Kiribati", in *Pacific Climate Cultures. Living Climate Chane in Oceania*, ed. by Tony Crook and Peter Rudiak-Gould (Warsaw: De Gruyter Open), pp. 21–33, https://doi.org/10.2478/9783110591415-003

Hodgson, Dorothy L. 2005. *The Church of Women. Gendered Encounters between Maasai and Missionaries*. Bloomington: Indiana University Press

——. 2011. "These Are Not Our Priorities. Maasai Women, Human Rights, and the Problem of Culture", in *Gender and Culture at the Limit of Rights*, ed. by Dorothy L. Hodgson (Philadelphia: University of Pennsylvania Press), pp. 138–57

Horton, Robin. 1975. "On the Rationality of Conversion (Part I)", *Africa*, 45.3: 219–35, https://doi.org/10.2307/1159632

Hulme, Mike. 2009. *Why We Disagree About Climate Change. Understanding Controversy, Inaction and Opportunity* (Cambridge, UK: Cambridge University Press), https://doi.org/10.1017/cbo9780511841200

——. 2014. "Foreword", in *How the World's Religions Are Responding to Climate Change*, ed. by Robin Globus Veldman, Andrew Szasz, and Randolph Haluza-DeLay (Abingdon: Routledge), pp. xii–iv

Igoe, Jim, and Dan Brockington. 1999. *Pastoral Land Tenure and Community Conservation. A Case Study from North-East Tanzania*, Pastoral Land Tenure Series 11 (London: International Institute for Environment and Development), https://pubs.iied.org/7385IIED/

Jasanoff, Sheila. 2010. "A New Climate for Society", *Theory, Culture & Society*, 27.2–3: 233–53. https://doi.org/10.1177/0263276409361497

Kempf, Wolfgang. 2017. "Climate Change, Christian Religion and Songs: Revisiting the Noah Story in the Central Pacific", in *Environmental Transformations and Cultural Responses. Ontologies, Discourses, and Practices in Oceania*, ed. by Eveline Dürr and Arno Pascht (New York: Palgrave Macmillan), pp. 19–48, https://doi.org/10.1057/978-1-137-53349-4_2

Kuruppu, Natasha, and Diana Liverman. 2011. "Mental Preparation for Climate Adaptation: The Role of Cognition and Culture in Enhancing Adaptive Capacity of Water Management in Kiribati", *Global Environmental Change*, 21.2: 657–69, https://doi.org/10.1016/j.gloenvcha.2010.12.002

Latour, Bruno. 1993. *We Have Never Been Modern* (Cambridge, MA: Harvard University Press)

Leslie, Paul, and J. Terrence McCabe. 2013. "Response Diversity and Resilience in Social-Ecological Systems", *Current Anthropology*, 54.2: 114–43, https://doi.org/ 10.1086/669563

Livingstone, David N. 2005. "Science, Text and Space: Thoughts on the Geography of Reading", *Transactions of the Institute of British Geographers*, 30.4: 391–401, https://doi.org/10.1111/j.1475-5661.2005.00179.x

Moser, Susanne C. 2010. "Communicating Climate Change. History, Challenges, Process and Future Directions", *Wiley Interdisciplinary Reviews: Climate Change*, 1.1: 31–53, https://doi.org/10.1002/wcc.11

——. 2016. "Reflections on Climate Change Communication Research and Practice in the Second Decade of the 21st Century. What More Is There to Say?", *Wiley Interdisciplinary Reviews: Climate Change*, 7.3: 345–69, https://doi.org/10.1002/wcc.403

Mosse, David. 2005. "Global Governance and the Ethnography of International Aid", in *The Aid Effect. Giving and Governing in International Development*, ed. by David Mosse and David Lewis (London: Pluto), pp. 1–36

Mosse, David, and David Lewis. 2006. "Theoretical Approaches to Brokerage and Translation in Development", *Development Brokers and Translators. The Ethnography of Aid and Agencies*, ed. by David Lewis and David Mosse (Bloomfield: Kumarian Press), pp. 1–26

Nerlich, Brigitte. 2017. "Science/Climate Communication: A View from Reception Theory", *University of Nottingham, Making Science Public Blog*, 28 September, https://blogs.nottingham.ac.uk/makingsciencepublic/2017/09/28/scienceclimate-communication-view-reception-theory/

Oba, Gufu. 2014. *Climate Change Adaptation in Africa. An Historical Ecology* (New York: Routledge)

Olson, David M., and Eric Dinerstein. 1998. "The Global 200. A Representation Approach to Conserving the Earth's Most Biologically Valuable Ecoregions", *Conservation Biology*, 12.3: 502–15, https://doi.org/10.1046/j.1523-1739.1998.012003502.x

Paton, Kathryn, and Peggy Fairbairn-Dunlop. 2010. "Listening to Local Voices. Tuvaluans Respond to Climate Change", *The International Journal of Justice and Sustainability*, 15.7: 687–98, https://doi.org/10.1080/13549839.2010.498809

Rayner, Steve. 1992. "Cultural Theory and Risk Analysis", in *Social Theories of Risk*, ed. by Dominic Golding and Sheldon Krimsky (Westport, CT: Praeger Publishers), pp. 83–115

——. 2003. "Domesticating Nature: Commentary on the Anthropological Study of Weather and Climate Discourse", in *Weather, Climate, Culture*, ed. by Sarah Strauss and Benjamin S. Orlove (Oxford, New York: Berg), pp. 277–90, https://doi.org/10.5040/9781474215947.ch-015

Rayner, Steve, and Clare Heyward. 2014. "The Inevitability of Nature as a Rhetorical Resource", in *Anthropology and Nature*, ed. by Kirsten Hastrup (New York, London: Routledge), pp. 125–46

Roe, Emery. 1999. *Except-Africa. Remaking Development, Rethinking Power* (New Brunswick, London: Transaction)

——. 1991. "Development Narratives, or Making the Best of Blueprint Development", *World Development*, 19.4: 287–300, https://doi.org/10.1016/0305-750X(91)90177-J

Rosengren, Dan. 2018. "Science, Knowledge and Belief. On Local Understandings of Weather and Climate Change in Amazonia", *Ethnos*, 83.4: 607–23, https://doi.org/10.1080/00141844.2016.1213760

Rottenburg, Richard. 2002. *Far-Fetched Facts. A Parable of Development Cooperation* (Cambridge, MA: MIT Press)

——. 2005. "Code-Switching, or Why a Metacode is Good to Have", in *Global Ideas. How Ideas, Objects and Practices Travel in the Global Economy*, ed. by Barbara Czarniawska and Guje Sevón (Malmö: Liber & Copenhagen Business School Press), pp. 259–75

Rubow, Cecilie, and Cliff Bird. 2016. "Eco-Theological Responses to Climate Change in Oceania", *Worldviews*, 20.2: 150–68, https://doi.org/10.1163/15685357-02002003

Rudiak-Gould, Peter. 2011. "Climate Change and Anthropology. The Importance of Reception Studies", *Anthropology Today*, 27.2: 9–12, https://doi.org/10.1111/j.1467-8322.2011.00795.x

——. 2013. *Climate Change and Tradition in a Small Island State. The Rising Tide*, Routledge Studies in Anthropology 13 (New York: Routledge), https://doi.org/10.4324/9780203427422

——. 2014. "Climate Change and Accusation", *Current Anthropology*, 55.4: 365–86, https://doi.org/10.1086/676969

Sachedina, Hassanali. 2008. "Wildlife is Our Oil. Conservation, Livelihoods and NGOs in the Tarangire Ecosystem, Tanzania" (PhD thesis, University of Oxford)

Sachedina, Hassanali, and Pippa Trench. 2009. "Cattle and Crops, Tourism and Tanzanite. Poverty, Land-Use Change and Conservation in Simanjiro District, Tanzania", in *Staying Maasai? Livelihoods, Conservation and Development in East African Rangelands*, ed. by Katherine Homewood, Patricia Kristjanson, and Pippa Trench (New York: Springer), pp. 263–98, https://doi.org/10.1007/978-0-387-87492-0_7

Sanders, Todd. 2004. "(En)Gendering the Weather. Rainmaking and Reproduction in Tanzania", in *Weather, Climate, Culture*, ed. by Sarah Strauss and Benjamin S. Orlove (Oxford: Berg), pp. 83–102, https://doi.org/10.5040/9781474215947.ch-005

Sheridan, Michael J. 2012. "Global Warming and Global War. Tanzanian Farmers' Discourse on Climate and Political Disorder", *Journal of Eastern African Studies* 6.2: 230–45, https://doi.org/10.1080/17531055.2012.669572

Spear, Thomas, and Richard Waller. 1993a. "Becoming Maasai, Being in Time", in *Being Maasai. Ethnicity and Identity in East Africa*, ed. by Thomas Spear and Richard Waller (London: James Currey), pp. 140–56

——. 1993b. *Being Maasai. Ethnicity and Identity in East Africa* (London: James Currey)

Spencer, Paul. 2003. *Time, Space, and the Unknown. Maasai Configurations of Power and Providence* (London, New York: Routledge)

Taddei, Renzo. 2012. "The Politics of Uncertainty and the Fate of Forecasters", *Ethics, Policy & Environment*, 15.2: 252–67, https://doi.org/10.1080/21550085.2012.685603

——. 2013. "Anthropologies of the Future: On the Social Performativity of (Climate) Forecasts", in *Environmental Anthropology. Future Trends*, ed. by Helen Kopnina (London: Routledge), pp. 244–63

Taylor, Anthony J. W. 1999. "Value Conflict Arising from Disaster", *The Australasian Journal of Disaster and Trauma Studies*, 2, https://www.massey.ac.nz/~trauma/issues/1999-2/taylor.htm

Taylor, Charles. 2007. *A Secular Age* (Cambridge, MA, London: The Belknap Press of Harvard University Press)

van Beek, Walter E. A. 2000. "Echoes of the End. Myth, Ritual and Degradation", *Focaal*, 35: 29–51

Waller, Richard. 1999. "They Do the Dictating and We Must Submit: The Africa Inland Mission in Maasailand", in *East African expressions of Christianity*, ed. by Thomas Spear and Isaria N. Kimambo (Oxford: James Currey), pp. 83-126

6. Living on the Frontier
Laypeople's Perceptions and Communication of Climate Change in the Coastal Region of Bangladesh

Shameem Mahmud

Despite an increased number of studies on public perceptions of climate change, little attention has been paid to the development of public understanding of climate change in developing and less-developed countries, which have contributed comparatively few greenhouse gas emissions. This chapter addresses this gap in the literature by exploring how people understand climate-change risks in an area at their forefront—the coastal region of Bangladesh. The study draws on in-depth interviews of local citizens and field observations. The interviews reveal a recurring theme of localizing climate-change risks in the context of local geohazards. Laypeople's personal exposure to extreme weather events, and experiences of seasonal variances, influence their interpretations of mediated and non-mediated climate change information. The risks of local geohazards are readily available as prior constructs in respondents' minds, and are intensified by newly acquired knowledge of climate change. The chapter concludes that laypeople's perceptions of climate-change impacts in Bangladesh are constructed on the basis of their place identity, on the one hand, and the frequency of regional geohazards, on the other.

https://doi.org/10.11647/OBP.0212.06

Introduction

Research on public perceptions of climate-change risks began in the late 1980s with the assumption that increased public understanding would help people make conscious livelihood decisions (Ockwell, Whitmarsh, and O'Neill 2009). Public perceptions have also been regarded as an important factor in making policy decisions to mitigate the risk of climate change (Leiserowitz 2006). The burgeoning climate change risk perception literature has identified paradoxical divergences between increased awareness and declining concern (Capstick et al. 2015; Poortinga et al. 2011; Whitmarsh 2011); misconceptions about the causes, impacts, and solutions to climate change (Leiserowitz and Smith 2010; Lorenzoni and Pidgeon 2006; Kempton 1997, 1991); scientific complexity and uncertainty (Etkin and Ho 2007; Weber 2006); perceptions of climate change as a spatially distant and temporally future risk (Whitmarsh and Upham 2013; Ockwell, Whitmarsh, and O'Neill 2009); and limited public behavioral responses to mitigate risks (Semenza et al. 2008; Lorenzoni, Nicholson Cole, and Whitmarsh 2007) despite different communication strategies (van der Linden 2014; Neverla and Taddicken 2011).

Despite a considerable increase in the number of studies exploring public understanding of climate change, few have examined the perceptions of people living in developing countries that are more vulnerable to (but less responsible for) climate change (Moser 2016, 2010). Previous studies can explain public understanding of climate change only in the countries largely perceived as responsible for the problem due to their greenhouse gas emissions (Neverla, Lüthje, and Mahmud 2012; Moser 2010). It is particularly interesting to explore public perceptions in countries where people are constantly exposed to regional geohazards such as storms, floods, and salinization that may increase and intensify as a consequence of climate change. This chapter addresses this research gap by investigating how the population of the coastal region of Bangladesh describes, explains, and understands the effects of climate change in the context of constant exposure to regional geohazards. It also investigates major information sources of climate change communication, which may affect laypeople's perceptions.

This research posits that, although climate change has emerged as a recent concern, the risks of regional geohazards are not new to the people living in this area. The effect of human-induced climate change will be to increase the likelihood or strength of such hazards. It is also reasonable to argue that increased communication about climate change, through journalistic media and other communication channels, may cause people living in these vulnerable areas to redefine their relationship with regional geohazards.

The chapter begins with a brief outline of Bangladesh's risks of natural hazards and vulnerability to climate change. Based on the extant literature, it then analyzes the role of respondents' geographic location and personal experience of natural hazards in influencing public perceptions of climate change. This review is followed by a description of the methodological design of the study. The research reveals a common social construct: the localization of climate-change risks within the context of local geohazards. Laypeople reveal that experiences of local hazards as well as personally observed weather and seasonal variances play important roles in their interpretations of mediated climate change information. This leads to the argument that laypeople's perceptions of climate-change impacts in the coastal region of Bangladesh are constructed on the basis of their place identity, on the one hand, and their experience of regional geohazards, on the other.

Bangladesh's Vulnerability to Climate Change

Although climate change poses global risks, these risks are not equally distributed around the world. Even different regions of a country may have different levels of vulnerability. Climate scientists have addressed this inequality by examining possible regional variability in the effects of climate change (Von Storch et al. 2011; Rannow et al. 2010; Rummukainen 2010). It has been widely documented that some developing countries may experience more severe impacts of global warming—in the forms of extreme weather, sea level rise, decreasing agricultural outputs, human health costs, and economic hardships—than developed countries (IPCC 2014; Mertz et al. 2009; UNFCCC

2007; Mirza 2003; Rosenzweig and Parry 1994; see also Chapter 7, Attribution Science).

Inequalities in the distribution of climate-change risks are exacerbated by limited hazard management capacities in developing countries (Adger et al. 2013; Nagel, Dietz, and Broadbent 2008; Ali 1999). Bangladesh holds a prominent position in the global imagination as being at the forefront of climate-change risks, given its geographic location, socio-economic characteristics, and natural hazard profile (Khan et al. 2011; Karim and Mimura 2008; Ali 1996). Many of its residents are experiencing the impacts of climate change (Coirolo et al. 2013); millions of people living in the country's coastal areas have been dubbed "climate victims" in political and civil society discourses (Walsham 2010; Friedman 2009).

Bangladesh is situated between two different, but mutually linked, geomorphological conditions. To the north, the Himalayan glaciers threaten to flood the country; more than 80% of its lands are floodplains (Brammer, Asaduzzaman, and Sultana 1996). To the south, the Bay of Bengal remains the source of notorious storms and storm surges that have made Bangladesh one of the most natural-hazard-prone countries in the world. Global warming is expected to increase the trends and frequencies of tropical storms (Schellnhuber 2007) in addition to raising the sea level, thereby affecting the country's densely populated coastal areas (Ahmed and Islam 2013). Residents of this area are frequently subjected to natural hazards in the form of tropical cyclones, floods, increased salinity in water and soil, storm surges and coastal erosion. Severe flooding that covers over 60% of the country occurs every four to five years, and a severe tropical cyclone hits, on average, every three years (MoEF 2009). Tropical cyclones in 1970 and 1991 are estimated to have killed 300,000 and 140,000 people, respectively.

Bangladesh is one of the few developing countries that has formulated a national climate change strategy plan (MoEF 2009). Discussions are ongoing to update the 2009 Bangladesh Climate Change Strategy and Action Plan (BCCSAP) to accommodate changing circumstances. The BCCSAP identifies the following risks of climate change in Bangladesh: (i) increased frequency of tropical cyclones; (ii) heavier and more erratic rainfall; (iii) lower and more erratic rainfall; (iv) sea level rise leading to the inundation of low-lying coastal areas displacing millions of people and paving the way for saline water intrusion; (v) warmer and more humid weather leading to an increased prevalence of disease; and (vi) increased intensity of drought in the country's northern and western areas. Vulnerability to these risks is further increased by the country's high population density. It is the world's eighth most populous (and one of the most densely populated) countries, with an estimated 162 million people living in an area of 147,570 km^2 (UNDP 2016). Around 70% of the population live in rural areas. Based on any development indicator, Bangladesh is a poor country. Its per capita income is US $1,080 per year (UNDP 2014), and more than 63 million people live below the poverty line (UNDP 2016). The overall adult literacy rate is only 58%. The economy is based on agriculture, which contributes about 22% to the country's GDP and employs 46% of its labour force (Bangladesh Bureau of Statistics 2014). Ready-made garment exports account for about 80% of total export earnings (Bangladesh Bureau of Statistics 2014).

Climate Change Risk Perception: The Role of "Distances" and "Experiences of Extreme Weather"

Public perceptions of and communication on climate change is a very complex research field. This complexity represents a double-edged sword. On the one hand, climate change communication shares features with other communication fields, most importantly, risk communication, health communication, and science communication. On the other hand, research on public perceptions of climate change is enriched by the contributions of various disciplinary traditions, such as human geography, and social and cognitive psychology (Nerlich, Koteyko, and Brown 2010). A detailed literature review of the field is beyond the scope of this chapter. Since one of the main objectives of the current research is to explore perceptions of the people living near natural hazards, this section focuses on research related to the role of *location or distance from the risk* (also see Chapter 4, Germany) and *personal experience of extreme weather* in constructing climate change risk perception.

Location or Distance from the Risk

Risk perception research widely demonstrates a correlation between the spatial proximity of risk and individuals' risk perceptions. For example, people living on the coastline or in flood-prone areas tend to express more concerns about the potential risks of natural hazards compared to those living in protected areas (Wester-Herber 2004). Similarly, people living close to nuclear power plants perceive the risk of nuclear hazards more seriously than those living further away (Venables et al. 2012), which clearly indicates the role of physical proximity to the origin of the risk. The notion of "proximity" or "distance", however, may also encompass temporal, social, and hypothetical dimensions. Trope and Liberman (2010), in their "construal-level theory of psychological distance", outline four key dimensions that might influence one's understanding and evaluation of risk events and issues: spatial or geographic distance, hypothetical or likelihood distance, temporal distance, and social distance.

The first dimension, spatial or geographic distance, is linked to the physical environment and location: people perceive climate-change risks as either (too) close to them or as something that will occur in more remote locations. Brody et al. (2008) and Kumar and Geneletti's (2015) studies on coastal inhabitants' climate change risk perceptions in the United States and India, respectively, found that people's physical location determines their level of concern about potential risk. There is also evidence that geographic proximity to the coast influences behavioral intentions to reduce carbon dioxide emissions on a personal level (Brody, Grover, and Vedlitz 2012). The role of spatial distance from the risk has been further demonstrated in studies examining the perception of natural hazards (Lorenzoni, Nicholson Cole, and Whitmatsh 2007; Weber 2006; Montz 1982), which have concluded that people living close to the geographic origin of hazards are more likely to take precautionary measures.

The second dimension, likelihood or hypothetical distance, relates to the probability that an event will occur (Liberman and Förster 2011). This can be compared to the "uncertainty" aspect of climate change. Existing research finds widespread public awareness and high levels of perceived concern since climate change emerged as a major policy issue in the late 1980s (Nisbet and Myers 2007; Lorenzoni and Pidgeon 2006). However, this high level of perceived concern was not stable; public concern about climate change has been found to fluctuate over this time. Jasanoff (2010) reported a widening gap between the public and experts in assessments of climate-change risks when laypeople were losing faith in climate science and showing a lower level of concern compared to other socio-economic risks. Other studies also report that while public awareness of (or familiarity with) climate change is increasing, concern about the issue has been fluctuating since the late 2000s, particularly in many developed countries (Capstick et al. 2015; Poortinga et al. 2011; Whitmarsh 2011). Pidgeon (2012) found that misleading media representations and the global financial crisis of 2008 and 2009 contributed to declining public concern about the issue. Some people also question whether climate change exists, and whether it is caused by human activity. Whitmarsh (2008b), for example, found that although the outright rejection of climate change among residents of the South of England was not widespread, a considerable number of people expressed some degree of uncertainty about its possible effects.

The third dimension that might influence people's understanding and evaluation of risk is "social" distances, which describe the extent to which people perceive the risks of climate change as applying to others rather than themselves. A number of studies drawing on cases from Europe and the United States have found that people generally perceive climate-change risks to be a global problem (Gifford et al. 2009; Uzzell 2000), which has serious implications for distant locations (Spence et al. 2011; Spence and Pidgeon 2010), where "other people" would be affected (see Chapter 2, Greenland). People in vulnerable regions, however, express more concern about the local impact of climate change (Dunlap, Gallup, and Gallup 1993).

Finally, Trope and Liberman's (2010) fourth dimension, temporal distance, refers to whether an event is likely to happen now or in the future. Temporal distance may explain whether the public perceives climate change as a current or future risk. According to previous studies, many inhabitants of developed countries do not perceive climate change as a significant personal risk because they think it will affect them far in the future, while it is more immediate for people living in geographically distant countries (Spence, Poortinga, and Pidgeon 2012; Leiserowitz

2006; Lorenzoni and Pidgeon 2006; Whitmarsh 2005; Bord, O'Connor, and Fisher 2000).

Experience of Extreme Weather

Climate change is a gradual and long-term phenomenon that is difficult to experience and observe in daily life. Yet most people describe concerns about climate change using evidence from personal experiences and observations (van der Linden 2014). Changes in weather patterns and extreme weather events are two possible impacts of climate change, which are often discussed in scientific and political arenas. These discourses are transferred to the public through news media and other communication channels. Eventually, laypeople recall their personal experiences of weather variances and natural hazard events when evaluating mediated knowledge of climate-change risks (Kempton 1997). Several studies support the notion that individuals' direct experiences and exposure to unusual weather events are important determinants of risk perception (see Chapter 2, Greenland; Lawrence, Quade, and Becker 2014; Lujala, Lein, and Rød 2014; Reser, Bradley, and Ellul 2014; Barnett and Breakwell 2001). Social psychologists describe this phenomenon as "the availability heuristic": personal experience of a recent natural hazard and heavy media coverage of it are likely to influence future risk evaluations (Slovic 1987; Tversky and Kahneman 1974).

Research Design

Study Location

This study was conducted in two adjacent unions[1]—Gabura and Padmapukur in Shyamnagar Upazila sub-district of the Satkhira district. Satkhira is a southwestern coastal district in Bangladesh located close to the world's largest mangrove forest—the Sundarbans. Both unions share similar characteristics in terms of the socio-economic conditions

---

1   A union is the lowest tier of the local government system in Bangladesh, consisting of a number of villages. At present there are 4,480 unions, which are run by directly elected representatives.

of their inhabitants, rural livelihood patterns, occupations, and vulnerability to natural hazards. Like other coastal regions, the villages in these unions are built on low-lying lands and are often protected by mud embankments. The area symbolizes human struggles against nature and the predators of the Sundarbans. Inhabitants of the area struggle with tigers and crocodiles when gathering resources from the mangrove forest, and are regularly exposed to natural hazards such as storms, storm surges, and intrusion of saltwater. The most notorious storms in the area were recorded in 1970, 1985, 1988, 1991, 2007, and 2009. While some affluent community members can afford houses made of brick and corrugated tin, most people in the study locations live in houses made of mud walls, bamboo and thatch, or roofs made of corrugated tin and thatch.

Fig. 6.1 A typical house structure made of local materials (photo by author, March 2012), CC-BY-4.0.

Fig. 6.2 Paddy fields are turned into saltwater shrimp enclosures (photo by author, March 2012), CC-BY-4.0.

For livelihood, people used to depend on rice cultivation, and resources of the Sundarbans and adjacent rivers (e.g., fish, crab, honey, and wood). However, in the last two decades, most of the agricultural lands have been transformed into shrimp enclosures given the higher profit margins. Similar to other rural areas of Bangladesh, nearly half of the population in the study area are unable to read and write. The poverty-struck villages are also characterized by a lack of infrastructure and poor health and education facilities, like many other coastal areas of Bangladesh.

Data Collection

Given the limited research on climate change perceptions in non-Western countries, data for this study were gathered using an inductive approach. The main data source was interviews, which were supplemented by prolonged field observations and notes. Interview participants were recruited following the strategy of theoretical sampling as suggested in Grounded Theory methodology (Charmaz 2006). I approached people for interviews who had comparatively more experiences of the social phenomena under study. A research guide, a local journalist, helped me find respondents and organize the initial interviews.

The first five interviewees included a schoolteacher (male), a local community leader (male), a shopkeeper (male), a non-governmental organization (NGO) field worker (female), and an elderly person (male). Once the initial interviews highlighted a general trajectory (identifying initial theoretical categories), the sampling technique was changed. In the second phase, the theoretical sampling technique was applied to saturate the missing lines of the identified theoretical categories with additional data. A total of thirty-eight people (twenty-six male and twelve female) were interviewed at their place of residence, which facilitated observation of the study location as well as the dynamic social relations of the study participants at the community level. The respondents included schoolteachers, a local community leader, NGO workers, shopkeepers, shrimp farmers, crab farmers, fishermen, and housewives. At least five held graduate degrees, three persons studied up to the twelfth grade, nine respondents studied between the sixth and tenth grades, four up to the fifth grade, and the remaining seventeen respondents were either illiterate or functionally literate.

There were three defined periods of data collection, which spanned from early 2011 to mid-2013. All the interviews were conducted in person in the local language (Bangla) and recorded using a digital audio recorder with the prior permission of the respondents. Then, the initial interviews were transcribed, with coding and analysis initiated with the first interviews, in accordance with Grounded Theory methodology techniques. The interview data were analyzed following the basic tenets of Grounded Theory, starting with open coding to focused coding, and ending in theoretical abstraction (Charmaz 2006) (see Table 6.1). The entire data analysis was conducted using the original Bangla transcriptions to avoid any distortion of the data due to translation. However, selected excerpts of the interviews were translated into English to present in this chapter.

Table 6.1 Perceived impacts of climate change (theoretical
abstraction from data)

Selective codes	Theoretical categories/ analytical memos	Emerging theoretical abstraction	Core category
"Increased and unprecedented storms"	*(I) Climate change as geohazards:* - Linking climate change to local events of geohazards. - Drawing a catastrophic image largely based on personal experiences and knowledge from secondary sources.	Localizing climate-change risks within the context of local geohazards.	Embedding cultural interpretations of climate-change impacts mainly derived from local hazard culture.
"Drowning in the sea"			
"Increased salinity"			
"Abrupt weather and seasonal cycles"	*(II) Climate change as weather and seasonal variances:* - Attributing meaning to climate change based on observed variances in local weather. - Comparing personal experiences of weather variances with mediated information.	Formally communicated information is evaluated against personal observations and memories.	
"Abrupt rainfall and increased temperature"			

Concrete → Abstract (top and bottom); left column label: Data/open codes

Findings

Two major patterns of public perceptions of climate-change impacts emerged from the interviews. First, the *regional (geo)hazard pattern* explains laypeople's understanding of climate-change impacts within the context of local geohazards. This research found three distinct social constructs that linked the impacts of climate change to local events of geohazards: "climate change as increased and unprecedented storms", "climate change as drowning in the sea", and "climate change means increased salinity". Second, the *weather and seasonal variances pattern* of climate-change impacts describes how the respondents attributed personal experiences of changes in local weather and seasons to climate change. These two patterns of perceptions, and examples of social constructs for each, are discussed in turn.

Climate Change as Regional (Geo)Hazards

Increased and Unprecedented Storms

The most common impact of climate change reported by respondents was storms—a frequent geohazard in the study area during the pre- and post-monsoon seasons. Respondents generally believed that storms and tidal floods in the region were caused by climate change, and that similar events would continue to increase in the future. They drew a causal link between storms and climate change, although climate science still struggles to clearly establish a link between a specific storm event and climate change (see Chapter 7, Attribution Science). This scientific uncertainty, however, did not prevent laypeople from making this link: "Climate change means to me storms. I've also heard about floods, rains, water and other disasters. But storms are the main" (Jaheda Akhtar, forty, F, illiterate, housewife).

A male shrimp farmer was similarly confident in the certainty of climate change and its impacts on the region: "Of course climate change is real. The reason I am saying this, actually we can see it by ourselves. It is already a reality for us. We are going through more storms and more [tidal] surges" (Avinash Mondol, forty-two, M, third grade, shrimp farmer). Likewise, a female NGO official supported the link between climate change and storms: "It's the storm that comes to my mind. There are other disasters, but storms are most important. I think other people have the same idea and they know it now that storms are results of climate change" (Shaheena Akhter, thirty-eight, F, MA, NGO worker).

Linking climate change and storms was so obvious that a number of respondents used the terms "climate" and "storms" synonymously while describing the possible impacts of climate change. This illustrates how strongly the natural hazard aspect of climate change is embedded in respondents' minds, likely due to respondents' physical proximity to previous storm events. For example: "Climate has been increased in recent years because of temperature rise. You can see it here. We didn't have such harsh climate in the past, but in recent times it has been increased. I mean the storms" (Gazi Mohhamad Ayub, thirty-two, M, ninth grade, unemployed).

The perceived link between climate change and storms was further demonstrated when some respondents explained that it was only the recent storm events and tidal floods that made it clear to them how "dreadful" the impacts of climate change would be. They described the effects as "immediate" and identified a temporal proximity to the risks of climate change. A female crab farmer said: "Honestly, it was not very clear to me when people talked about climate change. But, after the cyclone *Aila* now I can understand what it would like to be. *Aila* actually brought it [climate change] to us" (Maloti Mondol, forty-two, F, illiterate, crab farmer).

When respondents experience extreme weather events, they use their previous knowledge of climate change to attribute the event to this hitherto abstract phenomenon. When they were reminded that the geographic location of the area made it naturally susceptible to storms and tidal floods, most respondents agreed, but pointed to the "unprecedented" nature of recent storms and their "increased frequency", which compelled them to believe that the recent storms were linked to climate change. One elderly male respondent noted: "Of course you can say that. Storms have always been part of our life. But, nature of storms must have been changed. I have never experienced this kind of storm in my lifetime" (Moksed Ali, sixty-four, M, illiterate, unemployed). Another elderly respondent referred to changing patterns of storms as evidence of climate-change impacts. "Now they [storms] come without any sign. It's very difficult to predict nowadays" (Abul Hossain Mollah, seventy-nine, M, eighth grade, retired government employee). Another indicator was the perceived increased frequency of storms that many respondents attributed to climate change. "We didn't experience storms so frequently in the past... ... now what we heard about impacts of climate change" (Mohasheikh Sardar, forty-four, M, functional literacy, shrimp and agriculture farmer).

The respondents' descriptions clearly indicated that while storms had always impacted coastal people long before the debate on anthropogenic climate change, they believed that the rate and intensity of storms had recently started to increase. Persistent storms were one of the main motivations for drawing such perceptual constructs when respondents viewed the impacts of global climate change through the lens of local geohazards.

Drowning in the Sea

After increased storm events, "drowning in the sea" was the second-most widely cited social construct of climate change. Respondents believed that global warming-induced sea level rise would someday inundate large parts of Bangladesh's low-lying coastal areas. They often equated regular tidal surges with "sea level rise", and provided a number of personally observed changes that they believed were impacts of climate change. Some respondents gave detailed accounts of how global warming-led melting ice at the poles and mountains was affecting low-lying countries like Bangladesh. A shrimp farmer, for example, attributed the link between global warming and ice melting as follows: "I heard that ice is being melted because of rising temperature. Now, all the water from the melted ice is flowing downstream and flooding us" (SM Yunus Ali, fifty-five, M, illiterate, shrimp farmer).

Respondents obtained knowledge of "increased temperature", "ice melting", and "sea level rise" from a number of communication sources, such as media, NGOs, and local opinion leaders. Thereafter, the obtained knowledge was reconstructed through personal observations and experiences at the local level. Throughout the process, we observed a pattern of "localizing climate-change impacts" as a common social construct. As explained by a local community leader: "Let me give you an example. There are chars [river islands] near the Sundarbans. We have seen in the past that these chars didn't drown even in high tides. But now they totally disappear even in winter when the tides are not very high... what does it mean? Of course, the height of water has increased" (GM Rafiqul Islam, forty-one, M, twelfth grade, local community leader).

The attribution of "heightened water as an impact of climate change", however, was not prevalent in all respondents' descriptions. Apart from the impacts of climate change, some respondents cited "unplanned" water management at the local and regional levels as a major cause of flooding in the area. For example, shortly after attributing tidal surges to global warming, an elderly male respondent added: "I think some unplanned dikes and dams are also contributing to the rising water" (Abul Hossain Mollah, seventy, M, eighth grade, retired government employee).

A related prompt asked the respondent to elaborate, and he replied:

> This is my personal observation. In the last decades, a number of dikes and barrages have been built both in India and Bangladesh. These dikes have hampered the natural flow of rivers and created *chars* [river islands] with increased siltation ...How can the water flow to the sea when the rivers are filled up? Right? You see there is not enough space in the rivers to accommodate additional water (Abul Hossain Mollah, seventy, M, eighth grade, retired government employee).

The respondent then referred to the natural fight between fresh water and seawater, and how the seawater has been winning lately to intrude further into the country:

> In the past, the strength of the current in rivers was stronger than the high tides from the sea. Now, the situation is opposite. The strength of the high tides from the sea is winning against the current of river water because rivers do not have strong currents. As a result, seawater is coming more inside the lands (Abul Hossain Mollah, 70, M, eighth grade, retired government employee).

This respondent's observation gave an important insight into local causes of rising water in the rivers, an attribution that was shared by many other interviewees. A female NGO worker provided more insights into this "unplanned" water management issue:

> There was something like the green revolution in the 1986–87 for creating more croplands and irrigation. A lot of sluice gates were made at that time and a lot of dikes and dams to divert and manage river water for irrigation. They actually halted the natural flow of water. Because of these dikes and sluice gates, sediments cannot flow to the sea. They are being deposited in the riverbeds which ultimately decreases the water and sediment-carrying capacity of rivers Shaheena Akhter, thirty-eight, F, MA, NGO worker).

However, this sort of attribution did not overrule the climate change-induced sea level rise. Rather, respondents believed that the diversion of river water at the local and regional levels was escalating the siltation and tidal flooding for which climate change was equally blamed. A fisherman, for example, asserted: "You can say both—climate change and the unplanned dikes are causing floods" (Gazi Nur Uddin, forty-five, M, functional literacy, fishing and forestry).

The data clearly demonstrate that two different yet mutual social constructs prevail among residents of this area. One was grounded in their newly acquired knowledge of climate-change impacts: they believed the water levels had risen because of ice melting at the poles and in the Himalayan mountains. The second set of perceptions originates from the respondents' personal observations of the local water management system and sediment deposition in the riverbeds that they believed were detrimental to the natural water flow.

Increased Salinity

Salinity in water and soil is a common geohazard in the study area primarily because of its proximity to seawater. During every high tide, the seawater from the Bay of Bengal enters the coastal estuaries and flows up to one-hundred kilometers into the mainland through rivers and tributaries. In this process, seawater mixes with the fresh water of the downstream rivers of the Himalayas. The strength of tides from the sea and the current of downstream rivers play important roles in determining the level of salinity of water in the coastal region. Tidal surges of the sea tend to be more powerful than river currents. Respondents seemed to understand this natural process of salinization, but most seemed to be redefining their understanding of the salinity problem. They have been exposed to climate-change discourses recently through a number of communication channels (e.g., media, NGOs, and social contacts) in which salinization and sea level rise are often discussed as local impacts of climate change.

An NGO official involved in livelihood development programmes for poor people asserted:

> We have had some awareness programmes on coastal livelihood; we gave training to the villagers on how to survive in extreme saline conditions, for example, cultivating saline tolerant crops and vegetables. Climate change appears as a topic in these meetings as this is the main reason for the salinity problem in Gabura and other villages. I talked to many people and they now understand the real cause of salinity—it's global warming that brings saline water in this region (Khairul Alam, thirty-six, M, BA, NGO worker).

The influence of such awareness programmes was apparent in many respondents' descriptions. For example, a female respondent in her mid-twenties said: "Saline water is increasing because of climate change, that is what I heard from the organization's [NGO's] people" (Sonaban, twenty-five, F, illiterate, housewife).

A male respondent who worked as a crab farmer expressed a similar view when a related question reminded him about the area's susceptibility to salinity:

> I understand salinity is not a new problem, but the problem is being intensified. In the past, we could pump out fresh water from tube wells, but now all these wells are contaminated. Who's to blame for this? Climate change might be the reason. Particularly the organisation people are telling us that climate change is the main reason for salinity (Nipendranath Mondol, forty-five, M, twelfth grade, crab farmer).

However, many interviewees were not fully convinced that climate change was the only cause of the salinity, and were aware of the complex, inherently multi-causal nature of this problem. For example, a number of respondents identified shrimp aquaculture in former croplands as one of the root causes of increased salinization: "We should not always blame others for the problem. From my personal experience, shrimp enclosures are mainly contributing to the salinity problem" (Golam Mustafa Gazi, fifty-three, M, illiterate, honey hunter).

The shift towards shrimp farming in the early 1980s discussed above involved farmers introducing saline water from nearby rivers into their shrimp enclosures. This trend has had long-term consequences, including increased salinity in surface and groundwater as well as in the soil. An elderly male respondent asserted: "I think people are equally responsible for increased salinization. Owners of the *'ghers'* [shrimp enclosures] have brought in saline water from the rivers. To me, this is the main reason. Everything is salty now" (Abul Hossain Mollah, seventy, M, eighth grade, retired government employee).

Thus, while the interviewees initially referred to climate change as the main cause of the salinity problem, their detailed descriptions clearly identify geographic proximity and specific local economic and development activities such as shrimp farming as important contributors to the problem.

Climate Change as Weather and Seasonal Variances

Abrupt Seasonal Cycles

Interviewees overwhelmingly reported that the natural seasons in Bangladesh had been more abrupt and unexpected in recent years compared to the past, and they categorically blamed climate change for this shift. For many, climate change and unusual seasonal behavior were synonymous. A male crab farmer, for instance, describes climate change as follows: "To me, climate change means season change. Something that is not normal. It says change and change of something. Right? Something is not like before. I can see seasons are not like they were in the past" (Jeher Ali Mirza, fifty-five, M, illiterate, crab farmer).

Interviewees frequently mentioned that Bangladesh used to have six seasons, but now has only two or three that often overlap because of climate change. According to a male schoolteacher: "We used to have six seasons. But you will find them only in the books. In reality, there are two or three. Others are difficult to distinguish. But, this was not always the case; we were known as the country of six seasons" (Abul Bashar, thirty-six, M, MA, schoolteacher)

Similarly, a male shrimp farmer relied on the "past" or his "childhood" as the point of reference to describe changing seasonal patterns that he believed were a consequence of global climate change. "We hardly get enough rain during the monsoon. Summer is extremely hot and you will see changes in the winter season as well. These things are different than they used to be in the past" (Ashish Kumar Mondol, forty-eight, M, illiterate, shrimp farmer).

Erratic Rainfall and Temperature Anomalies

Erratic rainfall was a salient feature of respondents' descriptions of climate-change impacts at the local level. They often used terms such as "more erratic" or "more unpredictable" and "too early" or "delayed" to describe temporal variation in rainfall. Some respondents were so confident about the link between erratic rainfall and climate change that they recognized it as one of the most important pieces of evidence of climate change. Importantly, respondents frequently cited observations and experiences of their childhood or in the past to sustain their claims of erratic rainfall. In response to a question about the impacts of climate change, a boatman said:

> When it rains it pours endlessly for days incessantly. And when it doesn't rain, everything is dry. The problem is we don't get rain when we need, and we get it when actually we don't need it. I have heard this might be because of climate change. I am not sure though. But it seems to me true (Shomsher Ali, thirty-three, M, illiterate, boatman).

While a female respondent defined climate change as follows: "Well, climate change means [...] I think the weather is not the same as it was in the past. There are some visible changes in the weather pattern. To me, the rainy season has changed a lot" (Khadiza Banu, forty-five, F, functional literacy, housewife).

Other respondents take the monsoon or the rainy season as their point of reference in emphasizing links between erratic rainfall and climate change:

> We all knew that it would rain during the months of Ashar and Shrabon [Bengali months of the rainy season] and the month of Chaitra [Bengali month of the dry season] should be dry. That was the normal seasonal cycle. But, now all has been changed. Nowadays it is very uncertain (Moksed Ali, sixty-four, M, illiterate, unemployed).

Respondents also frequently attributed rising temperatures to climate change. A male schoolteacher, for example, explained:

We have been witnessing gradual temperature rise. It's too hot now. For example, my father built this house around 15 years ago. At that time, it was possible to stay inside the house during summer times. But, now, in the month of Chaitra it is too hot to stay inside the house. It feels like a hell. Extremely hot! (M Abul Bashar, thirty-six, M, M.A., schoolteacher).

Communicating Climate Change

Role of Media

Analysis of the interview data reveals two major sources of climate change knowledge: mass media and NGO advocacy programmes. As for mediated climate change knowledge, respondents mainly referred to radio and television. "It was on the television", "I heard it on radio", "there was a programme on television" and "mainly from television and radio news" were the sorts of assertions that respondents made while describing sources of climate change knowledge. An elderly respondent noted: "I heard about it [climate change] from the radio. It's a serious issue, very dangerous. Television also had some programmes on climate, but mainly radio" (Abul Hossain Mollah, seventy, M, eighth grade, retired government employee).

Similarly, a local community leader said: "I probably watched it on television news. You could find it almost every day at that time. Our prime minister was in a meeting [referring to COP 15 in Copenhagen] and it was a very big one on climate" (GM Rafiqul Islam, forty-one, M, twelfth grade, local community leader).

In the interview data, only three respondents, one schoolteacher and two NGO field workers, cited newspaper reports that had enhanced their knowledge of climate change. The internet and web- or mobile-based social networking sites were not mentioned during the interviews because of their complete absence regarding climate change communication in the study location.

Most people referred to television and radio while describing their understanding of and attitudes about climate change. "Rich countries' responsibility", "frequent storms", and "drowning in the sea" were some of the frames they learned about from the media which influenced their perceptions. Consider the following assertion: "A large part of Bangladesh will be drowned in the sea, the coastal area. I saw it on television" (Gazi Mohammad Ayub, thirty-two, M, ninth grade, unemployed).

As the data analysis progressed regarding the media's role in communicating climate change, habitual media use emerged as an important factor. In most cases, the media appeared to serve as a "passive source" for people, which indicated that climate change was not an interesting topic for them. Consider the following assertion: "It's not that I am very much interested about climate change and seek out there [in TV or radio]. I mainly listen to news, whatever it is, and find it [climate change] there" (Mohasheikh Sardar, forty-four, M, illiterate, shrimp-framing and agriculture).

Most respondents used the media primarily for entertainment. Only a handful of male respondents reported that they watched television or listened to radio news to stay up to date on contemporary national and global issues. A crab trader, for example, replied: "No, I didn't watch TV to know more about climate change. It was out there on the news" (Nipendranath Mondol, forty-five, M, twelfth grade, crab trader).

Role of NGOs

NGO advocacy programmes in the form of interpersonal and group communication emerged as the second most important channel of communicating information about climate change. NGOs' contribution to poverty alleviation, expanding education, women's empowerment, fighting for human rights and disaster management in Bangladesh is well recognized in the literature (Rahman 2006; Mercer et al. 2004). These organizations have served as important agents of socio-economic development and have focused their activities on the rural poor. More than 250 NGOs and civil society groups were listed in the local government office of the study area. These organizations were involved in a variety of projects including micro-credit to rural livelihood development, emergency support to cyclone victims, climate change adaptation, food security, mass education, forestation, and financing alternative livelihoods. After the cyclones in 2007 and 2009, many NGOs used foreign aid either to integrate a climate change component into their existing projects or to launch a separate programme to raise awareness. For example, a field worker for a local NGO described: "Climate change is now a priority focus of our existing programmes. We are working on mainstreaming climate change issues in providing livelihood supports and disaster management" (Mohsin Alam, M, thirty-one, MA, NGO field worker).

Thus, it is not surprising that NGOs were major sources of climate change knowledge for local residents. A female respondent asserted: "We didn't know too much about it [climate change] before the organization [NGO] people started to tell us" (Momtaj Begum, thirty, F, illiterate, housewife and fishing).

In addition to raising awareness, NGOs also communicated substantive messages related to the issue. A local shrimp farmer explained: "In the past, we didn't know reasons of storms and rising water. Now, we know; organization people told us it is because of climate change... Rich countries are liable for this" (Yunus Ali, M, fifty-five, illiterate, shrimp farmer).

Respondents also noted that NGOs used a participatory approach to engage people on different socio-economic issues, which enabled NGOs to deliver services to the target beneficiaries more effectively than government agencies. The focus on women, group formation and identifying opinion leaders in the community have important implications for climate change communication by NGOs. An NGO field worker described the importance of women in communicating their issues to the community:

> Our target beneficiaries are women. We give them loans. As part of it, they have to be members of a community group formed by us. We also provide training on health, family planning, disaster management, and alternative livelihood. Climate change has been included in our training programmes recently because it has been badly affecting people's lives (Mohsin Alam, thirty-one, M, MA, NGO field worker).

In Bangladesh, NGOs usually deliver their services to women after forming women's groups. These groups meet regularly to discuss issues related to their livelihood challenges. Usually, field workers of the NGOs deliver advocacy messages to the groups, which are then handed down to other members of the community through informal networks. An NGO field worker described her approach: "Normally, we meet at one of the member's *uthan* (courtyard) and it rotates to other members' houses" (Shahida Begum, thirty-five, F, twelfth grade, NGO field worker).

Discussion

The findings of this research clearly illustrate that the coastal people of Bangladesh widely believe that climate change is the underlying cause of their local geohazards. They also perceive climate change as both temporally (i.e., happening now) and spatially (i.e., happening here) close to them. This spatial and temporal proximity of climate-change risks has led people to personalize and localize the impacts.

Local residents frequently experience regional geohazards, and the media and NGOs often communicate these events as possible impacts of climate change. Thus, social constructs of climate-change impacts in this region appear to be constructed on the basis of place identity on the one hand, and availability heuristics through communication channels, on the other.

The "availability heuristics" concept infers that "a risk issue is likely to become powerful and capture the public's imagination if the cause, effect and victim are clearly identifiable" (Whitmarsh 2005), experienced or can be readily imagined (Kahneman, Slovic, and Tversky 1982). That is, recent personal experience of an extreme weather event or increased media coverage is likely to make people consider climate change a serious threat because it is readily available in their imagination. The risk perception literature supports this argument, as people living close to natural hazards tended to believe more strongly in the certainty of climate change (Brody et al. 2008). Leiserowitz (2006) also found that people who personally experienced environmental disasters tended to perceive climate change more emotionally compared to people who did not have such experiences (see also Chamila Roshani Perera and Rathnasiri Hewege 2013; Bulkeley 2000). By contrast, Whitmarsh (2005, 2008a) found that personal experiences of floods in the UK had little impact on the level of concern about climate change.

The findings of this research also demonstrate that, like people in many developed countries, the coastal people in Bangladesh fail to distinguish between the concepts of "climate" and "weather" when attributing evidence of climate change. They frequently cite personal observations of "abrupt weather" and "seasonal variances" to support the view that human-induced climate change is already taking place. This is in line with the studies of Read et al. (1994); Goebbert et al.

(2012); Capstick and Pidgeon (2014) and Taylor, Bruine de Bruin, and Dessai (2014), which all indicated laypeople's confusion in conceptually distinguishing between weather and climate (Etkin and Ho 2007; Kempton 1997; Read et al. 1994) when changing weather patterns were attributed to climate change (Capstick and Pidgeon 2014; Hulme 2014).

Radio and television were found to be the primary sources of mediated climate change knowledge for inhabitants of the study area. The media's role in communicating climate change issues from the domains of science and politics to the public has been well established (Nisbet 2009; Sampei and Aoyagi-Usui 2009; Cabecinhas, Lázaro, and Carvalho 2008; Weingart, Engels, and Pansegrau 2000a). In particular, Sampei and Aoyagi-Usui (2009) find that television and daily newspapers are the main sources of information about environmental and climate change for laypeople in Japan. However, the current research did not find that the print media played an important role in communicating climate change information in the study area. Low literacy rates, economic hardship, and geographic location have reduced public access to newspapers considerably. This implies that climate change communication strategies should take into account the local media environment and levels of access to different types of media.

We find that the study respondents believe climate change has been affecting them badly, but display little interest in seeking information on climate change from the media. Most interviewees reported consuming television and radio content for entertainment; climate change issues were generally perceived as either "too boring" or "too complicated". In rural Bangladesh, watching television or listening to the radio are largely regarded as leisure time activities. Rural people struggle with a number of hardships, and there are very limited options for entertainment. Media use thus serves as an escape from their daily troubles, as described in media use and gratifications theory (Blumler and Katz 1974; Katz, Blumler and Gurevitch 1974). These viewing patterns highlight the need for "infotainment" programmes on climate change issues.

Identifying the important role of interest groups, most notably NGOs, in communicating climate change information is one of the significant contributions of this research. Respondents' framing of climate change as linked to local geohazards and the politicization of the issue was informed by the NGOs. NGOs' role in politicizing the global climate change debate has received scholarly attention in recent years; they are important actors in mediating the issues to the public, policy makers and other stakeholders (e.g., Doyle 2009; Corell and Betsill 2001). However, the scientific literature has thus far overlooked NGOs' contribution in facilitating grassroots climate-change awareness. This study finds that NGOs are particularly successful at employing a participatory approach to communicate the issue to the public, and at creating community opinion leaders. In particular, they were able to create awareness of climate-change risks among women, who had comparatively limited access to the media.

Conclusion

This research explored public perceptions and the communication of climate-change risks in the coastal regions of Bangladesh, an area besieged by a number of regional geohazards that are worsening due to human-induced climate change. Previous research in this area has mainly been limited to examining cases in developed countries; the perceptions of people living on the "frontline" of climate change effects have been largely unexplored until now. Two main research questions guided this study. First, how do people in this area perceive risks of climate change? Second, how do they communicate these risks?

The study's findings reveal two important factors in constructing public perceptions: "local hazard culture" and "local communication environment". Regarding the former, public perceptions of climate-change risks are situated and constructed within the contexts of specific characteristics of local geohazards. Laypeople in the study area cited personal experiences of natural hazards and changes in local weather patterns as evidence of climate change. Accordingly, local events and their characteristics serve as important filters through which to interpret the risks of climate change.

The second factor, the local communication environment, highlighted the important role of respondents' sources of knowledge, notably the media (e.g., television) and formal (e.g., NGOs) and informal social contacts. The mutual relationship between these two factors means that people obtain climate change information from a number of sources, and this information is understood within a specific cultural framework of natural hazards.

Future research should address two additional areas. First, the study's methodological approach could be applied to other coastal regions of Bangladesh as well as hazard-prone regions in other countries. Such a cross-cultural comparison could explore the role of a specific hazard culture in the social construction of the concept of climate change. Second, future research should analyze NGOs' communication strategies, tools and content, as well as their role (and effectiveness) in creating awareness and motivating behavioral changes of laypeople.

References

Adger, W. Neil, Jon Barnett, Katrina Brown, Nadine Marshall, and Karen O'Brien. 2013. "Cultural Dimensions of Climate-change impacts and adaptation", *Nature Climate Change*, 3.2: 112–17, https://doi.org/10.1038/nclimate1666

Ahmed, Shamsun N., and Aminul Islam. 2013. "Equity and Justice Issues for Climate Change Adaptation in Water Resource Sector", in *Climate Change Adaptation Actions in Bangladesh*, ed. by Rajib Shaw, Fuad Mallick, and Aminul Islam (Tokyo: Springer Japan), pp. 143–63, https://doi.org/10.1007/978-4-431-54249-0_9,

Ali, Anwar. 1996. "Vulnerability of Bangladesh to Climate Change and Sea Level Rise through Tropical Cyclones and Storm Surges", *Water, Air, and Soil Pollution*, 92.1–2: 171–79, https://doi.org/10.1007/978-94-017-1053-4_16

——. 1999. "Climate-change Impacts and Adaptation Assessment in Bangladesh", *Climate Research*, 12: 109–16, https://doi.org/10.3354/cr012109

Bangladesh Bureau of Statistics. 2014. *Statistical Year Book Bangladesh 2014* (Dhaka: Bangladesh Bureau of Statistics (BBS), Statistics and Informatics Division (SID), Ministry of Planning, Government of the People's Republic of Bangladesh)

Barnett, Julie, and Glynis M. Breakwell. 2001. "Risk Perception and Experience: Hazard Personality Profiles and Individual Differences", *Risk Analysis*, 21.1: 171–78, https://doi.org/10.1111/0272-4332.211099

Blumler, Jay G., and Elihu Katz (eds). 1974. *The Uses of Mass Communications: Current Perspectives on Gratifications Research*, Sage Annual Reviews of Communication Research 3 (Beverly Hills, London: Sage)

Bord, Richard J., Robert E. O'Connor, and Ann Fisher. 2000. "In What Sense Does the Public Need to Understand Global Climate Change?", *Public*

Understanding of Science, 9.3: 205–18, https://doi.org/10.1088/0963-6625/9/3/301

Brammer, Hugh, Mohammed Asaduzzaman, and Parvin Sultana. 1996. "Effects of Climate and Sea-Level Changes on the Natural Resources of Bangladesh", in *The Implications of Climate and Sea-level Change for Bangladesh*, ed. by Richard A. Warrick and Qazi K. Ahmad (Dordrecht: Kluwer Academic Publishers), pp. 143–203, https://doi.org/10.1007/978-94-009-0241-1_4

Brody, Samuel, Himanshu Grover, and Arnold Vedlitz. 2012. "Examining the Willingness of Americans to Alter Behaviour to Mitigate Climate Change", *Climate Policy*, 12.1: 1–22, https://doi.org/10.1080/14693062.2011.579261

Brody, Samuel, et al. 2008. "A Spatial Analysis of Local Climate Change Policy in the United States: Risk, Stress, and Opportunity", *Landscape and Urban Planning*, 87.1: 33–41, https://doi.org/10.1016/j.landurbplan.2008.04.003

Bulkeley, Harriet. 2000. "Common Knowledge?: Public Understanding of Climate Change in Necastle, Australia", *Public Understanding of Science*, 9.3: 313–33, https://doi.org/10.1177/096366250000900301

Cabecinhas, Rosa, Alexandra Lázaro, and Anabela Carvalho. 2008. "Media Uses and Social Representations of Climate Change", in *Communicating Climate Change: Discourses, Mediations and Perceptions*, ed. by Anabela Carvalho (Braga: Centro de Estudos de Comunicação e Sociedade (CECS)), pp. 170–89

Capstick, Stuart, Lorraine Whitmarsh, Wouter Poortinga, Nick Pidgeon, and Paul Upham. 2015. "International Trends in Public Perceptions of Climate Change over the Past Quarter Century", *WIREs Climate Change*, 6.1: 35–61, https://doi.org/10.1002/wcc.343

Capstick, Stuart, and Nicholas Pidgeon. 2014. "Public Perception of Cold Weather Events as Evidence For and Against Climate Change", *Climatic Change*, 122.4: 695–708, https://doi.org/10.1007/s10584-013-1003-1

Chamila Roshani Perera, Liyanage, and Chandana Rathnasiri Hewege. 2013. "Climate Change Risk Perceptions and Environmentally Conscious Behaviour among Young Environmentalists in Australia", *Young Consumers*, 14.2: 139–54, https://doi.org/10.1108/17473611311325546

Charmaz, K. 2006. *Constructing Grounded Theory: A Practical Guide Through Qualitative Analysis* (London: Sage)

Coirolo, Cristina, Stephen Commins, Iftekharul Haque, and Gregory Pierce. 2013. "Climate Change and Social Protection in Bangladesh: Are Existing Programmes Able to Address the Impacts of Climate Change?", *Development Policy Review*, 31.2: o74-o90, https://doi.org/10.1111/dpr.12040

Corell, Elisabeth, and Michele M. Betsill. 2001. "A Comparative Look at NGO Influence in International Environmental Negotiations: Desertification and Climate Change", *Global Environmental Politics*, 1.4: 86–107, https://doi.org/10.4324/9781315092546-22

Doyle, J. 2009. "Climate Action and Environmental Activism: The Role of Environmental NGOs and Grassroots Movements in the Global Politics of Climate Change", in *Climate Change and the Media*, ed. by T. Boyce and J. Lewis (New York: Peter Lang Publishing, Inc), pp. 103–16.

Dunlap, Riley E., George H. Gallup, and Alec M. Gallup. 1993. "Of Global Concern: Results of the Health of the Planet Survey", *Environment*, 35.9: 7–39, https://doi.org/10.1080/00139157.1993.9929122

Etkin, David, and Elise Ho. 2007. "Climate Change: Perceptions and Discourses of Risk", *Journal of Risk Research*, 10.5: 623–41, https://doi.org/10.1080/13669870701281462

Friedman, Lisa. 2009. *How Climate Change is Making Refugees in Bangladesh*, https://www.scientificamerican.com/article/climate-change-bangladesh2/

Gifford, Robert, Leila Scannell, Christine Kormos, Lidia Smolova, Anders Biel, et al. 2009. "Temporal Pessimism and Spatial Optimism in Environmental Assessments: An 18-Nation Study", *Journal of Environmental Psychology*, 29.1: 1–12, https://doi.org/10.1016/j.jenvp.2008.06.001

Goebbert, Kevin, Hank C. Jenkins-Smith, Kim Klockow, Matthew C. Nowlin, and Carol L. Silva. 2012. "Weather, Climate, and Worldviews: The Sources and Consequences of Public Perceptions of Changes in Local Weather Patterns", *Weather, Climate, and Society*, 4.2: 132–44, https://doi.org/10.1175/wcas-d-11-00044.1

Hulme, Mike. 2014. "Attributing Weather Extremes to 'Climate Change': A Review", *Progress in Physical Geography*, 38.4: 499–511, https://doi.org/10.1177/0309133314538644

IPCC. 2014. *Climate Change 2014 Synthesis Report: Contribution of Working Groups I, II and III to the Fifth Assessment Report of the Intergovernmental Panel on Climate Change* (Geneva: IPCC), https://www.ipcc.ch/report/ar5/syr/

Jasanoff, Sheila. 2010. "A New Climate for Society", *Theory, Culture & Society*, 27.2–3: 233–53, https://doi.org/10.1177/0263276409361497

Kahneman, Daniel, Paul Slovic, and Amos Tversky. 1982. *Judgment and Uncertainty: Heuristics and Biases* (Cambridge, UK: Cambridge University Press)

Karim, Mohammed F., and Nobuo Mimura. 2008. "Impacts of Climate Change and Sea-Level Rise on Cyclonic Storm Surge Floods in Bangladesh", *Global Environmental Change*, 18.3: 490–500, https://doi.org/10.1016/j.gloenvcha.2008.05.002

Katz, Elihu, Jay G. Blumler, and Michael Gurevitch. 1974. "Uses and Gratifications Research", *Public Opinion Quarterly*, 37.4: 509, https://doi.org/10.1086/268109

Kempton, Willett. 1991. "Lay Perspectives on Global Climate Change", *Global Environmental Change*, 1.3: 183–208, https://doi.org/10.1016/0959-3780(91)90042-r

——. 1997. "How the Public Views Climate Change", *Environmental Science and Policy for Sustainable Development*, 39.9: 12–21

Khan, Aneire E., Wei Xun, Habibul Ahsan, and Paolo Vineis. 2011. "Climate Change, Sea-Level Rise, & Health Impacts in Bangladesh", *Environmental Science and Policy for Sustainable Development*, 53.5: 18–33

Kumar, Parveen, and Davide Geneletti. 2015. "How Are Climate Change Concerns Addressed by Spatial Plans?: An Evaluation Framework, and an Application to Indian cities", *Land Use Policy*, 42: 210–26, https://doi.org/10.1016/j.landusepol.2014.07.016

Lawrence, Judy, Dorothee Quade, and Julia Becker. 2014. "Integrating the Effects of Flood Experience on Risk Perception with Responses to Changing Climate Risk", *Natural Hazards*, 74.3: 1773–94, https://doi.org/10.1007/s11069-014-1288-z

Leiserowitz, Anthony. 2006. "Climate Change Risk Perception and Policy Preferences: The Role of Affect, Imagery, and Values", *Climatic Change*, 77.1–2: 45–72, https://doi.org/10.1007/s10584-006-9059-9

Leiserowitz, Anthony, and Nicholas Smith. 2010. *Knowledge of Climate Change across Global Warming's Six Americas* (New Haven: Yales Project on Climate Change Communication), http://environment.yale.edu/climate-communication-OFF/files/Knowledge_Across_Six_Americas.pdf

Liberman, Nira, and Jens Förster. 2011. "Estimates of Spatial Distance: A Construal Level Theory Perspective", in *Spatial Dimensions of Social Thought*, ed. by Thomas W. Schubert (Berlin: De Gruyter Mouton), pp. 109–29, https://doi.org/10.1515/9783110254310.109

Lorenzoni, Irene, Sophie Nicholson-Cole, and Lorraine Whitmarsh. 2007. "Barriers Perceived to Engaging with Climate Change among the UK Public and their Policy Implications", *Global Environmental Change*, 17.3–4: 445–59, https://doi.org/10.1016/j.gloenvcha.2007.01.004

Lorenzoni, Irene, and Nick F. Pidgeon. 2006. "Public Views on Climate Change: European and USA Perspectives", *Climatic Change*, 77.1-2: 73–95, https://doi.org/10.1007/s10584-006-9072-z

Lujala, Päivi, Haakon Lein, and Jan K. Rød. 2014. "Climate Change, Natural Hazards, and Risk Perception: The Role of Proximity and Personal Experience", *Local Environment*, 20.4: 489–509, https://doi.org/10.1080/13549839.2014.887666

Mercer, Alec, Mobarak Hossain Khan, Muhammad Daulatuzzaman, and Joanna Reid. 2004. "Effectiveness of an NGO Primary Health Care Programme in Rural Bangladesh: Evidence from the Management Information System",

Health Policy and Planning, 19.4: 187–98, https://doi.org/10.1093/heapol/czh024

Mertz, Ole, Kirsten Halsnaes, Jorgen E. Olesen, and Kjeld Rasmussen. 2009. "Adaptation to Climate Change in Developing Countries", *Environmental Management*, 43.5: 743–52, https://doi.org/10.1007/s00267-008-9259-3

Mirza, M. 2003. "Climate Change and Extreme Weather Events: Can Developing Countries Adapt?", *Climate Policy*, 3.3: 233–48, https://doi.org/10.1016/s1469-3062(03)00052-4

MoEF. 2009. *Bangladesh Climate Change Strategy and Action Plan, 2009* (Dhaka: Ministry of Environment and Forests, Govt. of the People's Republic of Bangladesh)

Montz, Burrell E. 1982. "The Effect of Location on the Adoption of Hazard Mitigation Measures", *The Professional Geographer*, 34.4: 416–23, https://doi.org/10.1111/j.0033-0124.1982.00416.x

Moser, Susanne C. 2010. "Communicating Climate Change: History, Challenges, Process and Future Directions", *Wiley Interdisciplinary Reviews: Climate Change*, 1.1: 31–53, https://doi.org/10.1002/wcc.11

——. 2016. "Reflections on Climate Change Communication Research and Practice in the Second Decade of the 21st Century: What More Is There to Say?", *Wiley Interdisciplinary Reviews: Climate Change*, 7.3: 345–69, https://doi.org/10.1002/wcc.403

Nagel, Joane, Thomas Dietz, and Jeffrey Broadbent. 2008. *Workshop on Sociological Perspectives on Global Climate Change: May 30–31, 2008* (Washington, DC: National Science Foundation), http://www.asanet.org/research/NSFClimateChangeWorkshop_120109.pdf

Nerlich, Brigitte, Nelya Koteyko, and Brian Brown. 2010. "Theory and Language of Climate Change Communication", *Wiley Interdisciplinary Reviews: Climate Change*, 1.1: 97–110, https://doi.org/10.1002/wcc.2

Neverla, Irene, Corinna Lüthje, and Shameem Mahmud. 2012. "Challenges to Climate Change Communication through Mass Media in Bangladesh: A Developing Country Perspective", in *Rethinking Climate Change Research: Clean-Technology, Culture and Communication*, ed. by Pernille Almlund, Per H. Jespersen, and Søren Riis (Farnham: Ashgate), https://www.researchgate.net/publication/282662399_Challenges_to_Climate_Change_Communication_Through_Mass_Media_in_Bangladesh_A_Developing_Country_Perspective

Neverla, Irene, and Monika Taddicken. 2011. „Klimawandel aus Sicht der Medienutzer: Multifaktorielles Wirkungsmodell der Medienerfahrung zur komplexen Wissensdomäne Klimawandel", *Medien & Kommunikationswissenschaft*, 59.4: 505–25, https://doi.org/10.5771/1615-634x-2011-4-505

Nisbet, Matthew C., and Teresa Myers. 2007. "The Polls Trends: Twenty Years of Public Opinion about Global Warming", *Public Opinion Quarterly*, 71.3: 444–70, https://doi.org/10.1093/poq/nfm031

Nisbet, Matthew C. 2009. "Communicating Climate Change: Why Frames Matter for Public Engagement", *Environment: Science and Policy for Sustainable Development*, 51.2: 12–23, https://doi.org/10.3200/envt.51.2.12-23

Ockwell, David, Lorraine Whitmarsh, and Saffron O'Neill. 2009. "Reorienting Climate Change Communication for Effective Mitigation: Forcing People to be Green or Fostering Grass-Roots Engagements?", *Science Communication*, 30.3: 305–27, https://doi.org/10.1177/1075547008328969

Pidgeon, Nick. 2012. "Public Understanding of, and Attitudes to, Climate Change: UK and International Perspectives and Policy', *Climate Policy*, 12.1: 85-106, https://doi.org/10.1080/14693062.2012.702982

Poortinga, Wouter, Alexa Spence, Lorraine Whitmarsh, Stuart Capstick, and Nick F. Pidgeon. 2011. "Uncertain Climate: An Investigation into Public Scepticism about Anthropogenic Climate Change", *Global Environmental Change*, 21.3: 1015–24, https://doi.org/10.1016/j.gloenvcha.2011.03.001

Rahman, Sabeel. 2006. "Development, Democracy and the NGO Sector", *Journal of Developing Societies*, 22.4: 451–73, https://doi.org/10.1177/0169796x06072650

Rannow, Sven, Wolfgang Loibl, Stefan Greiving, Dietwald Gruehn, and Burghard C. Meyer. 2010. "Potential Impacts of Climate Change in Germany: Identifying Regional Priorities for Adaptation Activities in Spatial Planning", *Landscape and Urban Planning*, 98.3–4: 160–71, https://doi.org/10.1016/j.landurbplan.2010.08.017

Read, Daniel, Ann Bostrom, M. Granger Morgan, Baruch Fischhof, and Tom Smuts. 1994. "What Do People Know About Global Climate Change?", *Risk Analysis*, 14.6: 959–70, https://doi.org/10.1111/j.1539-6924.1994.tb00066.x

Reser, Joseph P., Graham L. Bradley, and Michelle C. Ellul. 2014. "Encountering Climate Change: 'Seeing' is More than 'Believing'", *Wiley Interdisciplinary Reviews: Climate Change*, 5.4: 521–37, https://doi.org/10.1002/wcc.286

Rosenzweig, Cynthia, and Martin L. Parry. 1994. "Potential Impact of Climate Change on World Food Supply", *Nature*, 367.6459: 133–38, https://doi.org/10.1038/367133a0

Rummukainen, Markku. 2010. "State-of-the-Art Regional Climate Models", *Wiley Interdisciplinary Reviews: Climate Change*, 1.1: 82–96, https://doi.org/10.1002/wcc.8

Sampei, Yuki, and Midori Aoyagi-Usui. 2009. "Mass-Media Coverage, its Influence on Public Awareness of Climate-Change Issues, and Implications for Japan's National Campaign to Reduce Greenhouse Gas Emissions", *Global Environmental Change*, 19.2: 203–12, https://doi.org/10.1016/j.gloenvcha.2008.10.005

Schellnhuber, Hans Joachim. 2007. *Climate Change as a Security Risk* (London: Earthscan), https://doi.org/10.4324/9781849775939

Semenza, Jan C., David E. Hall, Daniel J. Wilson, Brian D. Bontempo, David J. Sailor, et al. 2008. "Public Perception of Climate Change Voluntary Mitigation and Barriers to Behavior Change", *American Journal of Preventive Medicine*, 35.5: 479–87, https://doi.org/10.1016/j.amepre.2008.08.020

Slovic, P. 1987. "Perception of Risk", *Science*, 236.4799: 280–85, https://doi.org/10.1126/science.3563507

Spence, Alexa, Wouter Poortinga, Christopher Butler, and Nick Pidgeon. 2011. "Perceptions of Climate Change and Willingness to Save Energy Related to Flood Experience", *Nature Climate Change*, 1.1: 46–49, https://doi.org/10.1038/nclimate1059

Spence, Alexa, and Nick Pidgeon. 2010. "Framing and Communicating Climate Change: The Effects of Distance and Outcome Frame Manipulations", *Global Environmental Change*, 20.4: 656–67, https://doi.org/10.1016/j.gloenvcha.2010.07.002

Spence, Alexa, Wouter Poortinga, and Nick Pidgeon. 2012. "The Psychological Distance of Climate Change", *Risk Analysis: An Official Publication of the Society for Risk Analysis*, 32.6: 957–72, https://doi.org/10.1111/j.1539-6924.2011.01695.x

Taylor, Andrea, Wändi Bruine de Bruin, and Suraje Dessai. 2014. "Climate Change Beliefs and Perceptions of Weather-Related Changes in the United Kingdom", *Risk Analysis*, 34.11: 1995–2004, https://doi.org/10.1111/risa.12234

Trope, Yaacov, and Nira Liberman. 2010. "Construal-Level Theory of Psychological Distance", *Psychological Review*, 117.2: 440–63, https://doi.org/10.1037/a0020319

Tversky, Amos, and Daniel Kahneman. 1974. "Judgment under Uncertainty: Heuristics and Biases", *Science*, 185.4157: 1124–31, https://doi.org/10.1017/cbo9780511809477.002

UNDP. 2014. *Human Development Report 2014: Sustaining Human Progress—Reducing Vulnerability and Building Resilience* (New York: UNDP), http://hdr.undp.org/en/content/human-development-report-2014

——. 2016. *Human Development Report 2016: Human Development for Everyone* (New York: UNDP), http://hdr.undp.org/sites/default/files/2016_human_development_report.pdf

UNFCCC. 2007. *Climate Change: Impacts, Vulnerabilities and Adaptation in Developing Countries* (Bonn: UNFCCC), http://unfccc.int/resource/docs/publications/impacts.pdf

Uzzell, David L. 2000. "The Psycho-Spatial Dimension of Global Environmental Problems", *Journal of Environmental Psychology*, 20.4: 307–18, https://doi.org/10.1006/jevp.2000.0175

van der Linden, S. 2014. "Towards a New Model for Communicating Climate Change", in *Understanding and Governing Sustainable Tourism Mobility: Psychological and Behavioural Approaches*, ed. by Scott A. Cohen et al. (London: Routledge), pp. 243–75

Venables, Dan, Nick F. Pidgeon, Karen A. Parkhill, Karen L. Henwood, and Peter Simmons. 2012. "Living with Nuclear Power: Sense of Place, Proximity, and Risk Perceptions in Local Host Communities", *Journal of Environmental Psychology*, 32.4: 371–83, https://doi.org/10.1016/j.jenvp.2012.06.003

Von Storch, Hans, and Jonas Bhend. 2011. "Regional Climate Services: Illustrated with Experiences from Northern Europe", *Zeitschrift Für Umweltpolitik & Umweltrecht*, 34: 1–15, https://doi.org/10.1007/s00382-007-0335-9

Walsham, Matthew. 2010. *Assessing the Evidence: Environment, Climate Change and Migration in Bangladesh* (Dhaka: IOM) http://publications.iom.int/system/files/pdf/environment_climate_change_bangladesh.pdf

Weber, Elke U. 2006. "Experience-Based and Description-Based Perceptions of Long-Term Risk: Why Global Warming Does Not Scare Us (Yet)", *Climatic Change*, 77.1–2: 103–20

Weingart, Peter, Anita Engels, and Petra Pansegrau. 2000. "Risks of Communication: Discourses on Climate Change in Science, Politics, and the Mass Media", *Public Understanding of Scienced*, 9.3: 261–83, https://doi.org/10.1088/0963-6625/9/3/304

Wester-Herber, M. 2004. "Underlying Concerns in Land-Use Conflicts: The Role of Place-Identity in Risk Perception", *Environmental Science and Policy*, 7.2: 109–16, https://doi.org/10.1016/j.envsci.2003.12.001

Whitmarsh, Lorraine. 2005. "A Study of Public Understanding of and Response to Climate Change in the South of England" (PhD thesis, University of Bath).

——. 2008a. "Are Flood Victims More Concerned about Climate Change than Other People?: The Role of Direct Experience in Risk Perception and Behavioural response", *Journal of Risk Research*, 11.3: 351–74, https://doi.org/10.1080/13669870701552235

——. 2008b. "What's in a Name?: Commonalities and Differences in Public Understanding of 'Climate Change' and 'Global Warming'", *Public Understanding of Science*, 18.4: 401–20

——. 2011. "Scepticism and Uncertainty about Climate Change: Dimensions, Determinants and Change over Time", *Global Environmental Change*, 21.2: 690–700, https://doi.org/10.1016/j.gloenvcha.2011.01.016

Whitmarsh, Lorraine, and Paul Upham. 2013. "Public Responses to Climate Change and Low-Carbon Energy", in *Low-Carbon Energy Controversies*, ed. by Roberts Thomas (Abingdon: Routledge), pp. 14–43

7. Extreme Weather Events and Local Impacts of Climate Change
The Scientific Perspective

Friederike E. L. Otto

While global and regional temperature increases are the most certain indicators of anthropogenic climate change, due to the emissions from burning fossil fuels, the damage caused by climate change is most clearly manifest in changes in seasons and extreme weather events. Recent advances in the attribution of extreme weather events, combined with newly available observations of past weather and climate, have made it possible to causally link high-impact extreme events to human-induced climate change. The level of confidence in these findings, however, varies according to the type of event and region of the world. While the increase in heatwaves can be quantified with confidence in most parts of the world, attribution assessments for droughts and hurricanes are much more uncertain.

https://doi.org/10.11647/OBP.0212.07

From Global Climate Change to Local Impacts

Scientific reports, political debates, and to a large extent the media, use global mean temperature rise as *the* metric to determine how humans are changing the climate by burning fossil fuels. Most prominently, the United Nations Framework Convention on Climate Change negotiations used this metric to measure and discuss the changing climate. More recently, the December 2015 Paris Agreement defines the central goal of international climate policy and politics as limiting global mean temperature increases to well below 2°C.

It is, however, not the abstract measure of global mean temperature that will cause loss and damage from climate change. Instead, the impacts of climate change primarily manifest themselves through rising sea levels and the changing risks of extreme weather events. This chapter provides a scientific account of the kinds of changes that are already being experienced locally, and the changes that we might see in the future. It then assesses what climate scientists know about the attribution of weather events (such as those discussed elsewhere in this volume) to global warming.

From Greenhouse Gas Emissions to Weather

The levels of scientific evidence and our ability to quantify each link in the chain of causality from greenhouse gas emissions via global mean temperature increases to local effects on weather and hydrology, to impacts affecting society, and loss and damage vary considerably (see Fig. 7.1). The evidence of the first element, from individual countries' (Skeie et al. 2017) and companies' (Heede 2014) greenhouse gas emissions to concentrations in the atmosphere, is unambiguous. The next link in the chain—from atmospheric CO_2 concentrations to global mean temperature—is also certain and has been known for a long time (IPCC 2014). Recent quantitative assessments estimate the global mean temperature increase resulting from anthropogenic greenhouse gas emissions from the beginning of the industrial revolution to the present at about 1°C (Haustein et al. 2017). The evidence is also strong for the increase in large-scale average temperatures; the absolute changes are very different, but temperatures are increasing in all inhabited regions of the world. For instance in Europe, annual mean temperatures have changed twice as fast as the global average (Hegerl et al. 2018; IPCC 2014). In recent years an increase in global precipitation that is expected in a warming climate has also been observed and attributed to human-induced greenhouse gas emissions (Marvel et al. 2017; IPCC 2014; Allen and Ingram 2002).

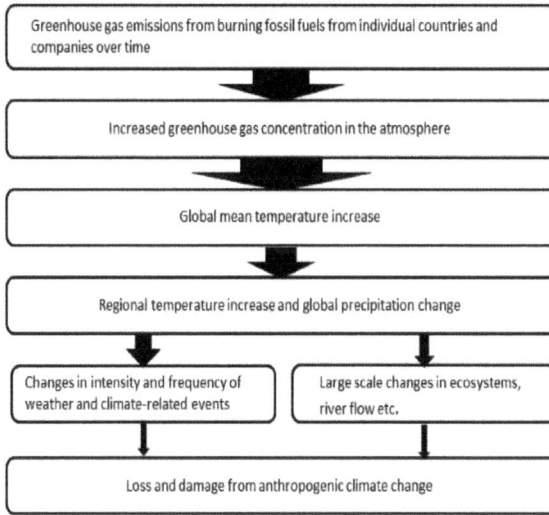

Fig. 7.1 Chain of causality from greenhouse gas emissions to loss and damage. The width of the arrow signifies the relative level of confidence in the quantitative assessments for different parts of the chain. For the last four boxes, confidence levels are very different depending on the region and type of event (e.g., heatwaves have high confidence, windstorms have low confidence) or impact (source: author, 2020).

The next—and arguably most crucial for the day-to-day lives of the majority of the world's population—link from anthropogenic climate change and global warming to local or regional individual weather and climate-related events has long been impossible to make with confidence. Yet, this has changed in recent years. One of the most noticeable local effects is the change in seasons: the mid-latitudes are experiencing an earlier spring and shorter winters, which is clearly caused by anthropogenic climate change (Santer et al. 2018; Christidis, Jones, and Stott 2014). Quantifying and establishing the link between individual weather events, which often cause significant damage, has been the focus of the emerging science of extreme event attribution (Otto 2017; Stott et al. 2016). This strand of research is ambitious because whether and to what extent an extreme weather event leads to impacts on natural and human systems depends not only on the meteorological hazard (i.e., the weather event) but also on the vulnerability and exposure to that hazard (i.e., who or what is in harm's way). Attributing damage and losses from extreme weather to climate change is thus possible with quantitative information on vulnerability and exposure and an understanding of local decision-making and governance structures, but very few studies have assessed this final link in the chain of causality (Mitchell et al. 2016; Schaller et al. 2016). However, enormous progress has been made in being able to robustly and routinely determine whether and to what extent anthropogenic climate change has altered the likelihood of extreme weather events to occur. Compared to only five years ago, we are now able to much more fully understand what anthropogenic global greenhouse gas emissions translate to for a particular location, because we can quantify the link between global warming and extreme weather. This is a crucial step that facilitates a true assessment of the changing risks on the level at which both individual and political decisions are made.

In 2003, a landmark publication (Allen 2003) suggested approaching the attribution question regarding the role of anthropogenic climate change in extreme weather events (Otto et al. 2016) in a probabilistic way. This approach entails assessing the likelihood of the extreme event in question to occur in the current climate, all man-made drivers included, and comparing it with the probability of its occurrence in a world without human-induced climate change in order to isolate and quantify the effect of climate change. The first paper to apply this approach was published in 2004; it found an increase of at least 100% in the likelihood that the 2003 European heatwave was attributable to human-induced climate change (Stott, Stone, and Allen 2004).

For today's climate, it is possible to use observations of weather and climate to estimate the likelihood that an event will occur. Yet we lack observations of a hypothetical, counterfactual world without anthropogenic climate change. Furthermore, we can only observe weather that has *occurred*; it is not possible to observe all weather events that are possible in a given climate. The method of event attribution thus relies on climate models to simulate possible weather, including the extreme event in question, in a given region and season accurately enough to draw robust conclusions on the role of climate change. Early studies applying the probabilistic event attribution approach employed a single climate model; thus, the results are heavily dependent on that model's reliability. A methodology has recently been developed that includes multiple models and several lines of evidence; a whole new field of climate science has emerged, and the methods are constantly improving. Two aspects of the methodology are important to stress here. First, the definition of an extreme event is a crucial part of the analysis and determines the outcome. In the most widely applied approach discussed here, the event is always defined as a type of weather that leads to an impact (e.g., extreme rainfall above a certain threshold in a particular area or season that causes flooding). Other approaches favor much more narrow definitions of an event by, e.g., conditioning on the exact atmospheric circulation state without assessing the likelihood of such a situation (Otto et al. 2016). The latter is useful for understanding the impact of climate change on weather but less useful for local decision-making. Second, attribution of extreme events is only reliable for the types of events for which climate models are available that realistically simulate the type of event; thus, the evidence of a causal link between anthropogenic climate change and localized weather events is not comprehensive.

Because the role of climate change is always mentioned by the media, by members of the public, and by people affected by the impacts of an extreme event, a significant amount of public attention has been focused on the development of the science. This exposure has also generated close scrutiny of scientific studies by peers; critical publications from both sides claim that scientists are too confident in their attribution statements (Bellprat and Doblas-Reyes 2016) or that they are too cautious (Lloyd and Oreskes 2018). While this might have produced temporary confusion in public understanding, overall, the public scientific discussions led to the comparably fast development of better and more robust methods of estimating changing hazards and reducing the uncertainty in the results.

In 2016, the US National Academy of Sciences asked climate scientists who were not involved in the science itself to write a report on the state of event attribution science and the robustness of the results of such studies. The resulting report (National Academies of Sciences, Engineering, and Medicine 2016) concluded that the results were robust overall, but not for all types of extreme events. For example, current climate models cannot reliably simulate very small-scale events, like flash floods, hail, or tornados, in order to assess the role of climate change in their occurrence. Nor do we have comprehensive observations on hailstorms with which to evaluate climate models. By contrast, since heatwaves and large-scale rainfall events are well observed and simulated, confidence in the attribution results is high. For droughts and other more complex events resulting from a combination of meteorological drivers (such as a lack of rainfall and high evaporation), confidence very much depends on which aspects of the droughts are studied. Despite these limitations, this independent report written by first-class climate scientists noted the thoroughness of the new approach to answering one of the most pressing questions when natural disasters strike. The report concluded that it is now possible to attribute individual weather events to anthropogenic climate change. It is thus only in the last few years that the role of climate change in extreme weather events around the world has been assessed; before, only a few studies examined specific events.

The advancement of event attribution science, and the availability of much improved simulations of future changes in extreme events, allow us to look at specific events and regions and thus enable a growing understanding of how climate change is impacting natural and human systems at the local level. This is necessary, as a warming climate has a twofold impact on local weather. First, given the increase in greenhouse gases in the atmosphere and rising global mean temperatures, we expect the number of heatwaves to increase and the coldwaves to decrease. Similarly, because a warmer atmosphere can hold more water vapor, extreme precipitation will increase globally, on average. Indeed, both of these impacts have already been observed (IPCC 2012).

Second, locally, however, this can look very different. The extent of extreme precipitation and temperature increases varies, because the thermodynamic (warming) effect is not the only way in which climate change affects weather and extreme events. Higher levels of greenhouse gases in the atmosphere, combined with more water vapor and a changing land surface, alter the atmospheric circulation, which affects where and when weather systems develop and how they progress. This so-called dynamic effect can be the same size as the thermodynamic effect or larger, and it can act in the same direction, thus increasing the risk of extreme precipitation, for example, more than would be expected from high temperatures alone (Risser and Wehner 2017; van Oldenborgh et al. 2017). The effects could also counteract each other, and thus prevent changes in extremes (Schaller et al. 2014). Thus, there are *a priori* three possible ways in which climate change affects weather and extreme weather; it can: (1) increase the likelihood of an event occurring, (2) decrease the likelihood of an event occurring, or (3) not affect the likelihood of events and extremes (Otto et al. 2016). Event attribution studies thus allow us to assess which of these three cases applies to a given type of weather event in a particular region and season and quantify the change.

While all three of these impacts can be the result of anthropogenic climate change alone, anthropogenic greenhouse gases in the atmosphere are not the only external driver affecting the likelihood of extreme events. For example, an increase in heat events can be offset by a cooling trend induced by implementing large-scale irrigation systems that keep the soil wet and thus reduce the land-atmosphere feedback (van Oldenborgh et al. 2018).

Aerosols of anthropogenic and natural origin also have a large influence on extreme weather events. Depending on the type of aerosol, they either reflect sunlight (and thus lead to a cooling) or absorb long-wave radiation (and thus exacerbate the effect of greenhouse gases). However, these are only the direct effects. Aerosols also interact with clouds in various ways, making it extremely hard to estimate their exact influence on the climate system. On a global scale they do lead to a cooling, however (IPCC 2014). In contrast to greenhouse gases, aerosols have a very short lifetime, on the order of days or weeks depending on how high in the atmosphere they are emitted. If the effect of greenhouse gases in an extreme event is masked by local aerosol pollution, irrigation or other land use changes, the observed data will reveal no trend in extremes.

Thus, if observations are the basis of decision-making, these can be fatally wrong. Increasing global attempts to combat air pollution will soon cancel the masking effect. This of course also means that attribution studies that focus only on the attribution of current extreme events without putting them in the context of future climate change miss an important part of the story. Very few studies have examined changing hazards across timescales from the past to the present and future. For instance, in the IPCC's fifth assessment report, detection and attribution (IPCC 2014) were discussed in a different chapter than projections of future climate change (IPCC 2014), and they used very different scales and methods of assessment. Future studies will address changing hazards, due to requests from city planners, disaster risk reduction experts, and other people involved in climate-change impacts and adaptation planning (van Aalst et al. 2018).

Thus, extreme event attribution allows us to assess the current impacts of climate change, in terms of weather as well as costs and ecological and societal implications; a comprehensive picture is slowly emerging, but there are huge gaps in the developing world. We also have the methodologies to combine this understanding with assessment of future climate change on geographical scales that matter for decision-making (e.g., Otto et al. 2018). However, while in theory it is possible to conduct an inventory of a large proportion of the changing hazards and extreme events taking place in a warming world, few studies have examined specific events in individual countries.

Regional Understanding

Bangladesh is in many ways the poster child of climate-change impacts (see Chapter 6, Bangladesh). On the one hand, its population is extremely vulnerable to flooding and saltwater intrusion (e.g. Warner and van der Geest 2013). Therefore, even small changes in the likelihood of extreme precipitation have a huge impact on the country's exposure to flooding. On the other hand, Bangladesh's scientists and government are acutely aware of this situation, and are exploring ways to deal with the impacts of climate change, including understanding and quantifying its impacts. For example, it was the first country to establish a national mechanism to address loss and damage from climate change—the costs (economic and otherwise) that mitigation and adaptation efforts have not avoided. While no articles have been published thus far that attribute individual extreme events in Bangladesh to anthropogenic climate change, a growing body of literature has confirmed that climate change increases the risk of damage from extreme events in a changing climate (Caesar et al. 2015; Ahmed, Diffenbaugh, and Hertel 2009; Karim and Mimura 2008).

Extreme weather events also occur elsewhere on the Indian subcontinent; each year it experiences droughts, floods, and heatwaves (Dash et al. 2007). In recent years many events have received high levels of international media attention because of heat records being broken (e.g., in Rajasthan in 2016, a large-scale heatwave in Andhra Pradesh in 2015) and large losses of assets (e.g., after massive flooding in Chennai in 2015), and of life (e.g., after exceptionally high monsoon rains in 2018). While there is little doubt that climate change plays a role in extreme weather events in India (Goswami et al. 2006), attribution studies of these recent events have shown that factors other than increased greenhouse gases continue to play a crucial role in driving extreme weather events over India (van Oldenborgh et al. 2018, 2016; Wehner at al. 2016). Natural variability as well as other man-made drivers like aerosol pollution play a particularly large role on small scales. At the same time, drivers outside the climate system, like river management, sewage water systems, and the sheer number of people in harm's way determine to a large degree the impacts of changing weather and climate. Understanding how these risks are changing throughout the world requires disentangling these factors and assessing their future trajectories.

In India and elsewhere, information on these changing risks is lacking at a scale where it is needed most—in the most vulnerable regions, where events cause the greatest impacts, where climate change is increasing risk, and where the media and the general public are asking questions about the root causes of disasters and their own vulnerability. India has a strong climate science community and widespread awareness of the fact that changes in the frequency and intensity of extreme weather events and rising sea levels caused by human-induced climate change threaten decades of development gains, and pose a clear and present danger to the social and economic welfare of communities. The city of Ahmedabad in Gujarat is a good example of how awareness of the changing risk of extreme heat has led to the development of a heatwave plan. The plan, consisting mostly of awareness rising and simple behavioral advice when heat wave warnings are being issued, dramatically reduced hospital admissions for heat-related conditions during the most recent heat wave compared with the 2005 heatwave (Knowlton et al. 2014). With heat-related diseases on the rise in a warming climate (Glaser et al. 2016), this example indicates that access to information on the extent to which extreme weather events are influenced by climate change and other factors can save lives.

The science base in Northern Europe for localized extreme events is broader: a large number of studies have assessed the changing risk of extreme events and heatwaves in particular. But the awareness and preparedness of individual cities is similar to those of India, although vulnerability and exposure are of course very different (cf. Chapter 4, Germany). The region's 2003 heatwave, which caused a reported 70,000 deaths (Donner, Muller, and Koppel 2015), demonstrated that preparedness is not a given in this part of the world (Aguiar et al. 2018). The event led the United Kingdom to develop a heatwave plan the following year, before the first attribution study (Stott, Stone, and Allen 2004) on this event was published. Subsequent heatwaves had less severe impacts, but all showed a clear attributable signal in likelihood and intensity to anthropogenic climate change (Christidis, Jones, and Stott 2014; Otto et al. 2012). Apart from heatwaves, an increase in extreme precipitation events is the most likely impact of anthropogenic climate change in Northern Germany. While there are no attribution studies on this specific area, given the similarity of the UK's climate, the attributable increase in the intensity and frequency of winter rainstorms there gives a good indication of what can be expected in Hamburg (Otto et al. 2018b; Schaller et al. 2016). In the summer, the risk of extreme precipitation is not currently changing (Otto et al. 2015a; Schaller et al. 2014). These results are in line with projections of future precipitation extremes, which show a significant increase in the winter but no clear sign of an attributable change in likelihood or intensity in the summer (IPCC 2012). Much less studied are the impacts of anthropogenic climate change on windstorms, which often cause casualties. The latest generation of regional climate models is good enough to enable realistic simulations, but the lack of comprehensive and reliable observational data on these events prevents assessments of this potential impact of climate change.

Our knowledge and understanding of extreme weather and their changes in a changing climate is again different in East Africa. Droughts and floods are the extreme events that have the largest impacts in the region; droughts have been studied intensely in recent years. Over the last few decades, an observed drying trend in much of the region has provided a puzzle for the climate science community, as the observations are at odds with model simulations that generally predict a wetting trend. Rowell, Booth, Nicholson, and Good (2015) have recently made progress in understanding this apparent paradox from a modelling perspective. In addition, a large number of event attribution studies (Liebmann et al. 2017; Philip et al. 2017; Uhe et al. 2017; Funk et al. 2016; Marthews et al. 2015; Otto et al. 2015a; Otto et al. 2015b) have helped to disentangle some of the drivers behind the observed trend and to identify the relative importance of different drivers in individual droughts in recent years.

The main finding of these event attribution studies is that anthropogenic climate change is not a key driver of the lack of rain in East Africa in recent years. Drought is a complex phenomenon; the higher global temperatures have exacerbated the drought by increasing the risk of dry soil moisture (Philip et al. 2017) through increased evaporation (Marthews et al. 2015), or via teleconnection through higher ocean temperatures (Funk et al. 2016). Teleconnections are patterns of variability in ocean temperatures; where they are analyzed for East Africa, the predominant mode of variability is the El Niño Southern Oscillation (Indeje, Semazzi, and Ogallo 2000). Variability in Indian Ocean temperatures is also important. Different modes in these patterns have always been crucial to determining seasonal rainfall and the likelihood of drought or flood in East Africa. A key question is whether (and to what extent) climate change will affect these patterns. There is currently insufficient evidence to draw definite conclusions, but some studies show that a warming trend, in combination with strong El Niño events, may increase the number of droughts (Funk et al. 2018) and flood disasters (Li et al. 2016) in the future.

In East Africa, huge advances in our understanding of rainfall variability and climate model reliability have been made possible by the better access to observed data (Funk et al. 2015), and close collaboration with scientists from East African Met Services and universities and international scientists. Yet many questions remain, particularly regarding the role of temperature.

An increase in temperature can contribute to drier soils and thus exacerbate the risk of drought, an effect often referred to when aiming to establish a climate change link (Funk et al. 2016). However, particularly in already arid regions, temperature often only exacerbates evaporation when moisture is present. Compounding the meteorological effects, it has been shown that hotter than normal temperatures have accelerated forage and water depletion across most of the region's pastoral and marginal agricultural areas (FEWS NET 2017), and thus identifying increasing temperatures as an additional stressor, which will have implications during future extreme heat and drought events, as well as for livelihood activities such as crop production (Republic of Kenya 2013).

East Africa has always experienced extreme weather events (see Chapter 5, Tanzania). But the fact that not all of the recent droughts were very extreme from a meteorological perspective highlights that it is high vulnerability that led to the severe societal impacts (Magnan et al. 2016).

In sum, the results generated by the nascent field of event attribution indicate that attribution studies around the world largely confirm our expectations—that a warming climate increases heatwaves and extreme rainfall events. The magnitude of this increase, however, varies widely by region and season, and local factors unrelated to climate change can prevent or decrease the risk of these extremes.

The examples above thus highlight that a heightened understanding of regional changes in individual types of extreme weather events facilitates preparation for all types of extreme weather. But if these advancements in natural science are not made relevant to those experiencing such weather and deciding how best to adapt, even perfect models and complete inventories of climate-change impacts will not prevent future loss and damage from extreme weather caused by climate change.

References

Aguiar, Francisca C., Julia Bentz, João M. N. Silva, Ana L. Fonseca, and Rob Swart. 2018. "Adaptation to Climate Change at Local Level in Europe: An Overview", *Environmental Science and Policy*, 86: 38–63, https://doi.org/10.1016/j.envsci.2018.04.010

Ahmed, Syud S., Noah S. Diffenbaugh, and Thomas W. Hertel 2009. "Climate Volatility Deepens Poverty Vulnerability in Developing Countries", *Environmental Research Letters*, 4.3: 34004, https://doi.org/10.1088/1748-9326/4/3/034004

Allen, Myles. 2003. "Liability for Climate Change", *Nature*, 421.6926: 891–92, https://doi.org/10.1038/421891a

Allen, Myles, and William J. Ingram. 2002. "Constraints on Future Changes in Climate and the Hydrologic Cycle", *Nature*, 419.6903: 224–32, https://doi.org/10.1038/nature01092

Bellprat, Omar, and Francisco Doblas-Reyes. 2016. 'Attribution of Extreme Weather and Climate Events Overestimated by Unreliable Climate Simulations", *Geophysical Research Letters*, 43.5: 2158–64, https://doi.org/10.1002/2015gl067189

Caesar, John, Tamara Janes, Amanda Lindsay, and Bhaski Bhaskaran. 2015. "Temperature and Precipitation Projections over Bangladesh and the Upstream Ganges, Brahmaputra and Meghna Systems", *Environmental Science: Processes and Impacts*, 17.6: 1047–56, https://doi.org/10.1039/c4em00650j

Christidis, Nikolaos, Gareth S. Jones, and Peter A. Stott. 2014. "Dramatically Increasing Chance of Extremely Hot Summers Since the 2003 European Heatwave", *Nature Climate Change*, 5: 46–50, https://doi.org/10.1038/nclimate2468

Dash, Sushil Kumar, Rajendra Kumar Jenamani, Shri S.R. Kalsi, and Subrat Kumar Panda. 2007. "Some Evidence of Climate Change in Twentieth-century India", *Climatic Change*, 85.3: 299–321, https://doi.org/10.1007/s10584-007-9305-9

Donner, Julie, Juliana M. Muller, and Johann Koppel. 2015. "Urban Heat: Towards Adapted German Cities?", *Journal of Environmental Assessment Policy and Management*, 17.2: 1550020, https://doi.org/10.1142/s1464333215500209

FEWS NET. 2017. "Kenya Food Security Outlook", *Early Warning of Food Insecurity*, https://reliefweb.int/sites/reliefweb.int/files/resources/KENYA_Food_Security_Outlook_February%202020_final_0.pdf

Field, Christopher B., Vicente Barros, Thomas F. Stocker, and Qin Dahe. 2012. *Managing the Risks of Extreme Events and Disasters to Advance Climate*

Change Adaptation (Cambridge, UK: Cambridge University Press), https://doi.org/10.1017/CBO9781139177245

Funk, Chris, Sharon E. Nicholson, Martin Landsfeld, Douglas Klotter, Pete Peterson, et al. 2015. "The Centennial Trends Greater Horn of Africa Precipitation Dataset", *Scientific Data*, 2: 150050, https://doi.org/10.1038/sdata.2015.50

Funk, Chris, Laura Harrison, Shraddhanand Shukla, Diriba Korecha, Tamuke Magadzire, et al. 2016. "Assessing the Contributions of Local and East Pacific Warming to the 2015 Droughts in Ethiopia and Southern Africa", *Bulletin of the American Meteorological Society*, 97.12: S75–S80, https://doi.org/10.1175/BAMS-D-16-0167.1

Glaser, Jason, Jay Lemery, Balaji Rajagopalan, Henry F. Diaz, Ramón Garcia-Trabanino, et al. 2016. "Climate Change and the Emergent Epidemic of CKD from Heat Stress in Rural Communities: The Case for Heat Stress Nephropathy", *Clinical Journal of the American Society of Nephrology*, 11.8: 1472–83, https://doi.org/10.2215/cjn.13841215

Goswami, Bhupendra Nath, V. Venugopal, D. Sangupta, M.S. Madhusoodanan, et al. 2006. "Increasing Trend of Extreme Rain Events Over India in a Warming Environment", *Science*, 314.5804: 1442–45, https://doi.org/10.1126/science.1132027

Haustein, Karsten, Myles Allen, Piers Forster, Friederike Otto, Daniel M. Mitchell, et al. 2017. "A Real-Time Global Warming Index', *Scientific Reports*, 7.1: 15417, https://doi.org/10.1038/s41598-017-14828-5

Heede, Richard. 2014. "Tracing Anthropogenic Carbon Dioxide and Methane Emissions to Fossil Fuel and Cement Producers, 1854–2010", *Climatic Change*, 122.1: 229–41, https://doi.org/10.1007/s10584-013-0986-y

Hegerl, Gabriele C., Stefan Brönnimann, Andrew Schurer, and Tim Cowan. 2018. "The Early 20th Century Warming: Anomalies, Causes, and ConsequencesW, *WIREs Clim Change*, 9.4: e522, https://doi.org/10.1002/wcc.522

Indeje, Matayo, Fredrick H. M. Semazzi, and Laban J. Ogallo. 2000. "ENSO Signals in East African Rainfall Seasons", *International Journal of Climatology*, 20.1: 19–46, https://doi.org/10.1002/(SICI)1097-0088(200001)20:1%3C19::AID-JOC449%3E3.0.CO;2-0

IPCC. 2012. *Managing the Risks of Extreme Events and Disasters to Advance Climate Change Adaptation: Special Report of the IPCC* (Cambridge, UK: Cambridge University Press) https://www.ipcc.ch/report/managing-the-risks-of-extreme-events-and-disasters-to-advance-climate-change-adaptation/

——. 2014. *Climate Change 2014 Synthesis Report: Contribution of Working Groups I, II and III to the Fifth Assessment Report of the Intergovernmental Panel on Climate Change* (Geneva: IPCC), https://www.ipcc.ch/report/ar5/syr/

Karim, Mohammed F., and Nobuo Mimura. 2008. "Impacts of Climate Change and Sea-Level Rise on Cyclonic Storm Surge Floods in Bangladesh",

Global Environmental Change, 18.3: 490–500, https://doi.org/10.1016/j.gloenvcha.2008.05.002

Knowlton, Kim, Suhas P. Kulkarni, Gulrez Shah Azhar, Dileep Mavalankar, Anjali Jaiswal, et al. 2014. "Development and Implementation of South Asia's First Heat-Health Action Plan in Ahmedabad (Gujarat, India)", *International Journal of Environmental Research and Public Health*, 11.4: 3473–92, https://doi.org/10.3390/ijerph110403473

Li, Chan-juan, Yuan-qing Chai, Lin-sheng Yang, and Hai-rong Li 2016. "Spatio-Temporal Distribution of Flood Disasters and Analysis of Influencing Factors in Africa", *Natural Hazards*, 82.1: 721–31, https://doi.org/10.1007/s11069-016-2181-8

Liebmann, Brant, Ileana Bladé, Chris Funk, Dave Allured, Xiao-Wei Quan, et al. 2017. "Climatology and Interannual Variability of Boreal Spring Wet Season Precipitation in the Eastern Horn of Africa and Implications for Its Recent Decline", *Journal of Climate*, 30.10: 3867–86, https://doi.org/10.1175/jcli-d-16-0452.1

Lloyd, Elisabeth A., and Naomi Oreskes. 2018. "Climate Change Attribution: When Is It Appropriate to Accept New Methods?", *Earth's Future*, 6.3: 311–25, https://doi.org/10.1002/2017ef000665

Magnan, K., Lisa Schipper, Maxine Burkett, Sukaina Bharwani, et al. 2016. "Addressing the Risk of Maladaptation to Climate Change", *WIREs Climate Change*, 7.5: 646–65, https://doi.org/10.1002/wcc.409

Marthews, Toby, Friederike E. L. Otto, Daniel Mitchell, Simon James Dadson, and Richard G. Jones. 2015. "The 2014 Drought in the Horn of Africa: Attribution of Meteorological Drivers", *Bulletin of the American Meteorological Society*, 96.12: 83–88, https://doi.org/10.1175/bams-d-15-00115.1

Marvel, Kate, Michela Biasutti, Céline Bonfils, Karl E. Taylor, Yochanan Kushnir, et al. 2017. "Observed and Projected Changes to the Precipitation Annual Cycle", *Journal of Climate*, 30.13: 4983–95, https://doi.org/10.1175/jcli-d-16-0572.1

Mitchell, Daniel, Clare Heaviside, Sotiris Vardoulakis, Chris Huntingford, Giacomo Masato, et al. 2016. "Attributing Human Mortality during Extreme Heat Waves to Anthropogenic Climate Change", *Environmental Research Letters*, 11.7: 74006, https://doi.org/10.1088/1748-9326/11/7/074006

National Academies of Sciences, Engineering, and Medicine. 2016. *Attribution of Extreme Weather Events in the Context of Climate Change* (Washington, DC: The National Academies Press), https://doi.org/10.17226/21852

Otto, Friederike E. L. 2017. 'Attribution of Weather and Climate Events', *Annual Review of Environment and Resources*, 42.1: 627–46, https://doi.org/10.1146/annurev-environ-102016-060847

Otto, Friederike E. L., Neil Massey, Geert Jan van Oldenborgh, Richard G. Jones, and Myles R Allen. 2012. "Reconciling Two Approaches to Attribution of the 2010 Russian Heat Wave", *Geophysical Research Letters*, 39.4, https://doi.org/10.1029/2011gl050422

Otto, Friederike E. L., Suzanne M. Rosier, Myles R. Allen, Neil R. Massey, Cameron J. Rye, et al. 2015a. "Attribution Analysis of High Precipitation Events in Summer in England and Wales over the Last Decade", *Climatic Change*, 132.1: 77–91, https://doi.org/10.1007/s10584-014-1095-2

Otto, Friederike E. L., Emily Boyd, Richard G. Jones, Rosalind J. Cornforth, Rachel James, et al. 2015b. "Attribution of Extreme Weather Events in Africa: A Preliminary Exploration of the Science and Policy Implications", *Climatic Change*, 132.4: 531–43, https://doi.org/10.1007/s10584-015-1432-0

Otto, Friederike E. L., Geert Jan van Oldenborgh, Jonathan Eden, Peter A. Stott, David J. Karoly, et al. 2016. "The Attribution Question", *Nature Climate Change*, 6.9: 813–16, https://doi.org/10.1038/nclimate3089

Otto, Friederike E. L., Karin van der Wiel, Geert Jan van Oldenborgh, Sjouke Philip, Sarah F. Kew, et al. 2018. "Climate Change Increases the Probability of Heavy Rains in Northern England/Southern Scotland Like Those of Storm Desmond—a Real-Time Event Attribution Revisited', *Environmental Research Letters*, 13.2: 24006, https://doi.org/10.1088/1748-9326/aa9663

Philip, Sjoukje, Sarah F. Kew, Geert Jan van Oldenborgh, Friederike Otto, Sarah O'Keefe, et al. 2017. "Attribution Analysis of the Ethiopian Drought of 2015", *Journal of Climate*, 31.6: 2465–86, https://doi.org/10.1175/jcli-d-17-0274.1

Republic of Kenya. 2013. *National Climate Change Action Plan 2013–2017* (Nairobi: Ministry of Environment and Mineral Resources), https://cdkn.org/wp-content/uploads/2013/03/Kenya-National-Climate-Change-Action-Plan.pdf

Risser, Mark D., and Michael F. Wehner. 2017. "Attributable Human-Induced Changes in the Likelihood and Magnitude of the Observed Extreme Precipitation during Hurricane Harvey", *Geophysical Research Letters*, 44.24: 12,457–64, https://doi.org/10.1002/2017gl075888

Rowell, David, Ben Booth, Sharon Nicholson, and Peter Good. 2015. "Reconciling Past and Future Rainfall Trends over East Africa", *Journal of Climate*, 28.24: 9768–88, https://doi.org/10.1175/JCLI-D-15-0140.1

Santer, Benjamin D., Stephen Po-Chedley, Mark D. Zelinka, Ivana Cvijanovic, Céline Bonfils, et al. 2018. "Human Influence on the Seasonal Cycle of Tropospheric Temperature", *Science*, 361.6399: 245-56, https://doi.org/10.1126/science.aas8806

Schaller, Nathalie, Friederike E. L. Otto, Geert Jan Van Oldenborgh, Neil R. Massey, Sarah Sparrow, et al. 2014. "The Heavy Precipitation Event of May–June 2013 in the Upper Danube and Elbe Basins", *American Meteorological Society*, 95.9: S69–S72, https://pdfs.semanticscholar.org/b7f3/90dab80a8363a3ef415e601df5a63800b6ac.pdf

Schaller, Nathalie, Alison L. Kay, Rob Lamb, Neil R. Massey, Geert Jan van Oldenborgh, et al. 2016. "Human Influence on Climate in the 2014 Southern England Winter Floods and Their Impacts", *Nature Climate Change*, 6.6: 627–34, https://doi.org/10.1038/nclimate2927

Skeie, Ragnhild B., Jan Fuglestvedt, Terje Bernsten, Glen P. Peters, Robbie Andrew, et al. 2017. "Perspective Has a Strong Effect on the Calculation of Historical Contributions to Global Warming", *Environmental Research Letters*, 12.2: 24022, https://doi.org/10.1088/1748-9326/aa5b0a

Stott, Peter A., Nikolaos Christidis, Friederike E. L. Otto, Ying Sun, Jean-Paul Vanderlinden, et al. 2016. "Attribution of Extreme Weather and Climate-Related Events", *WIREs Climate Change*, 7.1: 23–41, https://doi.org/10.1002/wcc.380

Stott, Peter A., Dáithí A. Stone, and Myles R. Allen. 2004. "Human Contribution to the European Heatwave of 2003", *Nature*, 432.7017: 610–14, https://doi.org/10.1038/nature03089

Uhe, Peter, Sjoukje Philip, Sarah Kew, Kasturi Shah, Joyce Kimutai, et al. 2017. "Attributing Drivers of the 2016 Kenyan Drought", *International Journal of Climatology*, 38: e554–68, https://doi.org/10.1002/joc.5389

van Aalst, Maarten, Richard Jones, Allan Lavell, Hans-Otto Pörtner, Debra Roberts, et al. 2018. *Bridging Climate Science, Policy and Practice. Report of the International Conference on Climate Risk Management, Pre-Scoping Meeting for the IPCC Sixth Assessment Report* (The Hague: IPCC), https://www.climatecentre.org/downloads/files/RCCC%20IPCC%20Nairobi%20Report%202018-4%20V5.pdf

van Oldenborgh, Geert J., Friederike E. L. Otto, Karten Haustein, and Krishna Achuta Rao. 2016. "17. The Heavy Precipitation Event of December 2015 in Chennai, India", *Bulletin of the American Meteorological Society*, 97.12: 87–90, https://doi.org/10.1175/bams-d-16-0129.1

van Oldenborgh, Geert J., Karin van der Wiel, Antonia Sebastian, Roop Singh, Julie Arrighi, et al. 2017. "Attribution of Extreme Rainfall from Hurricane Harvey, August 2017", *Environmental Research Letters*, 12.12: 124009, https://doi.org/10.1088/1748-9326/aa9ef2

van Oldenborgh, Geert J., Sjoujke Philip, Sarah Kew, Michiel van Weele, Peter Uhe, et al. 2018. "Extreme Heat in India and Anthropogenic Climate Change", *Natural Hazards and Earth System Sciences*, 18.1: 365–81, https://doi.org/10.5194/nhess-18-365-2018

Warner, Koko, and Kees van der Geest. 2013. "Loss and Damage from Climate Change: Local-Level Evidence from Nine Vulnerable Countries", *International Journal of Global Warming*, 5.4: 367–86, https://doi.org/10.1504/ijgw.2013.057289

Wehner, Michael, Dáithí Stone, Hari Krishnan, Krishna Achuta Rao, and Federico Castillo. 2016. "The Deadly Combination of Heat and Humidity in

India and Pakistan in Summer 2015", *Bulletin of the American Meteorological Society*, 97.12: S81–S86, https://doi.org/10.1175/bams-d-16-0145.1

List of Illustrations

Chapter 3

Chapter 4

Chapter 5

Chapter 6

Chapter 7

Index

About the Team

Alessandra Tosi was the managing editor for this book.

Adele Kreager performed the copy-editing and proofreading.

Anna Gatti designed the cover using InDesign. The cover was produced in InDesign using the Fontin font.

Melissa Purkiss typeset the book in InDesign and produced the paperback and hardback editions. The text font is Tex Gyre Pagella; the heading font is Californian FB.

Luca Baffa produced the EPUB, MOBI, PDF, HTML, and XML editions — the conversion is performed with open source software freely available on our GitHub page (https://github.com/OpenBookPublishers).

This book need not end here...

Share

All our books — including the one you have just read — are free to access online so that students, researchers and members of the public who can't afford a printed edition will have access to the same ideas. This title will be accessed online by hundreds of readers each month across the globe: why not share the link so that someone you know is one of them?

This book and additional content is available at:

https://doi.org/10.11647/OBP.0217

Customise

Personalise your copy of this book or design new books using OBP and third-party material. Take chapters or whole books from our published list and make a special edition, a new anthology or an illuminating coursepack. Each customised edition will be produced as a paperback and a downloadable PDF.

Find out more at:

https://www.openbookpublishers.com/section/59/1

Like Open Book Publishers

Follow @OpenBookPublish

Read more at the Open Book Publishers **BLOG**

Global Communications

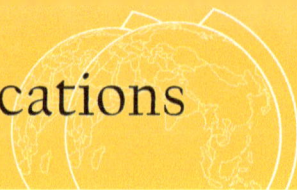

About the series

Global Communications is a book series that looks beyond national borders to examine current transformations in public communication, journalism and media. Books in this series will focus on the role of communication in the context of global ecological, social, political, economic, and technological challenges in order to help us understand the rapidly changing media environment. We encourage comparative studies but we also welcome single case studies, especially if they focus on regions other than Western Europe and North America, which have received the bulk of scholarly attention until now.

Empirical studies as well as textbooks are welcome. Books should remain concise and not exceed 300 pages but may offer online access to a wealth of additional material documenting the research process and providing access to the data. The series aspires to publish theoretically well-grounded, methodologically sound, relevant, novel research, presented in a readable and engaging way. Through peer review and careful support from the editors of the series and from the editorial team of Open Book Publishers, we strive to support our authors in achieving these goals.

Global Communications is the first Open Access book series in the field to combine the high editorial standards of professional publishing with the fair Open Access model offered by OBP. Copyrights stay where they belong, with the authors. Authors are encouraged to secure funding to offset the publication costs and thereby sustain the publishing model, but if no institutional funding is available, authors are not charged fees. Any publishing subvention secured will cover the actual costs of publishing and will not be taken as profit. In short: we support publishing that respects the authors and serves the public interest.

You can find more information about this serie at:
https://www.openbookpublishers.com/section/100/1

www.ingramcontent.com/pod-product-compliance
Lightning Source LLC
Chambersburg PA
CBHW040147270326
41929CB00025B/3414